AF773392

Studies in Systems, Decision and Control

Volume 119

Series editor

Janusz Kacprzyk, Polish Academy of Sciences, Warsaw, Poland
e-mail: kacprzyk@ibspan.waw.pl

About this Series

The series "Studies in Systems, Decision and Control" (SSDC) covers both new developments and advances, as well as the state of the art, in the various areas of broadly perceived systems, decision making and control- quickly, up to date and with a high quality. The intent is to cover the theory, applications, and perspectives on the state of the art and future developments relevant to systems, decision making, control, complex processes and related areas, as embedded in the fields of engineering, computer science, physics, economics, social and life sciences, as well as the paradigms and methodologies behind them. The series contains monographs, textbooks, lecture notes and edited volumes in systems, decision making and control spanning the areas of Cyber-Physical Systems, Autonomous Systems, Sensor Networks, Control Systems, Energy Systems, Automotive Systems, Biological Systems, Vehicular Networking and Connected Vehicles, Aerospace Systems, Automation, Manufacturing, Smart Grids, Nonlinear Systems, Power Systems, Robotics, Social Systems, Economic Systems and other. Of particular value to both the contributors and the readership are the short publication timeframe and the world-wide distribution and exposure which enable both a wide and rapid dissemination of research output.

More information about this series at http://www.springer.com/series/13304

Guidong Zhang · Bo Zhang
Zhong Li

Designing Impedance Networks Converters

Springer

Guidong Zhang
Guangdong University of Technology
Guangzhou
China

Zhong Li
FernUniversität in Hagen
Hagen, Nordrhein-Westfalen
Germany

Bo Zhang
South China University of Technology
Guangzhou, Guangdong
China

ISSN 2198-4182 ISSN 2198-4190 (electronic)
Studies in Systems, Decision and Control
ISBN 978-3-319-63654-2 ISBN 978-3-319-63655-9 (eBook)
DOI 10.1007/978-3-319-63655-9

Library of Congress Control Number: 2017946922

© Springer International Publishing AG 2018
This work is subject to copyright. All rights are reserved by the Publisher, whether the whole or part of the material is concerned, specifically the rights of translation, reprinting, reuse of illustrations, recitation, broadcasting, reproduction on microfilms or in any other physical way, and transmission or information storage and retrieval, electronic adaptation, computer software, or by similar or dissimilar methodology now known or hereafter developed.
The use of general descriptive names, registered names, trademarks, service marks, etc. in this publication does not imply, even in the absence of a specific statement, that such names are exempt from the relevant protective laws and regulations and therefore free for general use.
The publisher, the authors and the editors are safe to assume that the advice and information in this book are believed to be true and accurate at the date of publication. Neither the publisher nor the authors or the editors give a warranty, express or implied, with respect to the material contained herein or for any errors or omissions that may have been made. The publisher remains neutral with regard to jurisdictional claims in published maps and institutional affiliations.

Printed on acid-free paper

This Springer imprint is published by Springer Nature
The registered company is Springer International Publishing AG
The registered company address is: Gewerbestrasse 11, 6330 Cham, Switzerland

Guidong Zhang dedicates the book to his parents and his fiancée, Miss. Yanjun Wang.

Preface

Power electronics is concerned with control power conversion between various forms and/or characteristics to suit particular applications, and to realize the high-efficiency energy conversion, which plays a key role in, especially, renewable energy. New power electronics converters should be designed to better meet the ever higher demand for renewable energy applications. Power electronics refer to three aspects, i.e. power switches, converter topologies, and control strategies; while the topology is the most important issue, since the other two aspects are based on a specific topology.

In terms of a converter's input source being either a voltage source or a current source, traditional converters fall into two categories: voltage-source and current-source converters. It is, however, known that voltage-source converters suffer from the shoot-through problem, the inapplicability to inductive loads, and a limited output-voltage gain; while current-source converters have the open-circuit problem, the inapplicability to capacitive loads, and a limited output-current gain, which have greatly hindered the application of conventional converters in modern energy, especially renewable energy. In order to solve these problems, the concept of Z-source has been proposed in 2003, where Z stands for impedance, which is a general source, taking voltage source and current source as two extreme cases.

Since then, various Z-source based converters have been coined for specific applications. The design is, however, dependent on experts' experience, lacking a systematic designing methodology. Therefore, the main purpose of this book is first to understand impedance networks of nonlinear switch circuits and impedance networks matching, which involves input impedance matching, output impedance matching, and load phase matching. Taking the three matchings together into account leads to a set of criteria for designing an impedance source converter, which is to replace the traditional tedious, manual and experience-dependent design methods.

We hope that this book can serve as a teaching material for graduates and guidelines for engineers as designing an impedance source converter.

Guangzhou, China — Guidong Zhang
Guangzhou, China — Bo Zhang
Hagen, Germany — Zhong Li
May 2017

Acknowledgements

For the successful completion of this book, Guidong Zhang is indebted to Profs. Yun Zhang and Si-Zhe Chen for their fruitful discussions; Prof. Dongyuan Qiu for her helpful advice; Kai Chen, Lei Xu, Xuqin Liao, Shaoqing Zhou and Gengpeng Zhang. Furthermore, his greatest gratitude goes to his parents and grandparents for their forever love and spiritual support, and to his sister and brothers for their contributions to family.

The authors thank Alexander von Humboldt Research Group Linkage Programme (grant No. 2.4-IP-DEU/1009882), the National Natural Science Foundation of China (U1501251, 51307025) and the Foundation for Distinguished Young Talents in Higher Education of Guangdong (Grant No. 2016KQNCX039) for their financial support.

Guangzhou, China — Guidong Zhang
Guangzhou, China — Bo Zhang
Hagen, Germany — Zhong Li

Contents

Chapter 1
A Brief History of Power Electronics Converters

This chapter is to retrospect the development of power electronics converters, point out the challenges faced especially in renewable energy applications, and introduce a promising novel Z-source converter.

1.1 Power Electronics

Thunder, lightning and electric fish, which are related to electricity [1], were just natural phenomena in ancient time, and treated as myths but not energy in human life until the discovery of electrostatic phenomena by Thales of Miletus in 640-540 BC [2, 3]. Much later in 1752, B. Franklin discovered electricity [4] and in 1820 H.C. Örsted revealed electromagnetism [5]. Since then, a series of great discoveries on the principles of electricity and magnetism have been done by Volta, Coulomb, Gauss, Henry, Faraday, and others, leading to many inventions such as the batteries (1800), generators (1831), electric motors (1831), telegraphes (1837), and telephones (1876), to name just a few. Until the early 19th century the electrical science has been established and in the late 19th century the greatest progress has been witnessed in electrical engineering [6].

In 1882, the first power grid, which was a direct-current (DC) distribution system invented by T. Edison, was set up in New York to provide 110 V DC power to supply over 1,000 bulbs in a short distance. Then, the problem was how to transfer energy at a low loss from power plants to customers over a long distance through transmission lines [7]. It is now well known that electricity must be transmitted at high voltages and in the form of alternative-current (AC) because DC voltage cannot be increased or decreased by DC systems at that time [8]. In 1885, L. Gaulard and J.D. Gibbs developed a device, named a transformer, which can increase or decrease the electrical voltage of AC systems. Thereafter, G. Westinghouse applied the transformer in AC distribution systems to make the electricity available to be

© Springer International Publishing AG 2018

G. Zhang et al., *Designing Impedance Networks Converters*, Studies in Systems, Decision and Control 119, DOI 10.1007/978-3-319-63655-9_1

transmitted over long distances efficiently, which has promoted the development of electrical engineering [9].

Transformers played a vital rôle in electricity transmission, especially in the energy conversion between different voltages. However, transformers can only increase or decrease AC voltage (AC-AC) at the same frequency. Moreover, energy loss of transformers, magnetic radiations, huge volume, and high economic cost of copper limited their wider applications [10]. In practical applications, electric energy was expected to convert from one form to another, for instance, between AC and DC, or just to different voltages or frequencies, or some combinations of those, which cannot be fully fulfilled by transformers. To this end, novel techniques were required to solve those problems. With the developments of semiconductor switches technology, power electronics has come into being [11].

Power electronics refers to electric power, electronics and control systems. Electric power deals with the static and rotating power equipments for the generation, transmission and distribution of electric power; while electronics is concerned with solid state semiconductor power devices and circuits together with specified control systems for power conversion to meet the desired control objectives [12]. Power electronics is one of the main technologies to realize energy conversion with high efficiency. It is known that about 70% electric energy should be converted first by power electronics before it comes to our life and industries. Nowadays, power electronics has become a fundamental and critical technology in the development of energy conservation, especially for renewable energy [13–16].

The history of power electronics is linked up with the break-through and evolutions of power semiconductor devices [17]. The first power electronics device was the mercury arc rectifier developed in 1900, followed by other power devices, like metal tank rectifier, grid controlled vacuum tube rectifier, ignitron, phanotron, thyratron and magnetic amplifier, developed and deployed gradually for power control applications until 1950s. The second electronics revolution began in 1958 with the development of the commercial grade thyristor by the General Electric company (GE), indicating the beginning of a new era of power electronics. Since 1975, more turn-off power semiconductor elements have been developed and implemented in the next 20 years, which had vastly improved modern electronics. Included here are improved bipolar transistors (with fine structures, and also shorter switching times), metal-oxide-semiconductor field effects transistors (MOSFETs), gate turnoff thyristors (GTOs) and insulated gate bipolar transistors (IGBTs). Correspondingly, many different types of power semiconductor devices and power conversion techniques have been designed and proposed. The power electronics revolution has endowed us the ability to convert, shape and control power [18].

With the development of semiconductor devices, different kinds of control strategies have also been developed to realize specified purposes. For example, high-accuracy and high-frequency control methods based on chips like DSP, FPGA and CPLD were applied to meet the desired requirements and gain better control of the load; more accurate mathematical modeling methods of power converters enabled to gain better output features, reduce the energy loss and increase the efficiency; and improved control algorithms were utilized to improve the efficiency and robustness, reduce the complexity, and gain better output features.

1.2 Power Electronics Converters

Power electronics converters fall into four categories, i.e. AC-DC, AC-AC, DC-DC, and DC-AC converters, and they have been invented and found a wide spectrum of applications in, for instance, the transportation (electric/hybrid electric vehicles, electric locomotives, electric trucks), utilities (line transformers, generating systems, grid interface for alternative energy resources like solar, wind, and fuel cells, and energy storage), industrial/commercial (motor drive systems, electric machinery and tools, process control, and factory automation), consumer products (air conditioners/heat pumps, appliances, computers, lighting, telecommunications, uninterruptible power supplies, and battery chargers), and medical equipments. Moreover, with the advanced power electronics converters, high-voltage direct current (HVDC) systems are also available to replace some long-distance AC transmission systems with unique features. Nowadays, power electronics has become a scientific discipline [19].

With rapid development of modern industry, more severe problems are faced by power electronics: how to meet the requirements of the load; how to improve the efficiency and reliability of power semiconductor devices; how to design converters with smaller volume, less weight, and lower cost; how to reduce the number of power switches and thus the design complexity of converters and improve the robustness of the whole system; and how to minimize negative influence on other equipments in the electric power systems and on the electromagnetic environment [20].

In order to solve these problems, some advances were witnessed in the semiconductor switches of power converters, for example, integrated gate-commutated thyristors (IGCT) were invented to have lower conduction loss compared to the traditional high capacities switches. Accordingly, control strategies were also improved [21].

To design a new power electronics converter, one can, on the one hand, develop a new control strategy; on the other hand, one can design a novel power converter topology, so as to obtain specific outputs, more simple control, higher efficiency, less complexity, lower weight, minimal cost, and better robustness. In fact, a control strategy is specified to a certain topology, and the topology determines the control system. Therefore, it is of great significance to coin optimal power converter topologies to fulfill various requirements in applications.

1.3 Z-source Converters

In terms of the input source of a converter being either a voltage source or a current source, various traditional converters fall into two categories: voltage source and current source converters. It is, however, known that voltage source converters suffer from shoot-through problems, inapplicability to inductive loads, and limited output voltage gains; while current source converters have open-circuit problems, inapplicability to capacitive loads, and limited output current gains [22].

In order to solve these problems, a Z-source converter was firstly proposed by Peng in 2002 [23], by coupling an LC impedance network (a two-port network with a combination of two basic linear energy storage elements, i.e. L and C) with the DC source to form a novel source, named Z-source, which is a kind of impedance source (an impedance is denoted by Z) [24–26]. Impedance source can be regarded as a general source, including the current and the voltage sources as two extreme cases, i.e. an impedance source can be regarded as a current source when its equivalent impedance tends towards infinity; while as a voltage source when its equivalent impedance is equal to zero.

Since then, many novel impedance source converters with various topologies have been coined, such as quasi-Z-source converters, trans-Z-source converters, embedded-Z-source converters, which have been widely applied in wind energy systems [27–29], motor drives [30–33], vehicle systems [34–38], and solar energy systems [39–48]. Unfortunately, the design of a specific impedance source converter is still an art, lacking of a systematic design methodology, which hinders the extensive application of impedance source converters in practice.

It is remarked that designing an impedance source converter should be subject to the impedance network matching, which instructs how an impedance network is matched to the sources. Therefore, the essential impedance matching should be then introduced and clarified.

Thereafter, a profound analysis of voltage and current sources converters is to be carried out in order to explain why impedance source converters have the unique features over traditional converters firstly. Furthermore, the impedance networks matching is to be explained, which leads to a systematic methodology of designing impedance networks converters for various specific applications. Finally, some novel impedance source converters will be introduced for special industrial applications.

References

1. P. Moller, B. Kramer, Review: electric fish. BioSci. (Am. Inst. Biol. Sci.). **44**(11), 794–800, (1991)
2. J. Stewart, Intermediate electromagnetic theory. BioSci. (World Sci.). p. 50 (2001). ISBN 981-02-4471-1
3. B. Simpson, Electrical Stimulation and the Relief of Pain. Elsevier Health Sciences. pp. 6–7. ISBN 0-444-51258-6, 2003

4. J. Srodes, *Franklin: The Essential Founding Father*, (Regnery Publishing, 2002), pp. 92–94. ISBN 0-89526-163-4, 2002
5. R.S. Kirby, *Engineering in History*, (Courier Dover Publications, 1990), pp. 331–333. ISBN 0-486-26412-2
6. J.F. Keithley, *The Story of Electrical and Magnetic Measurements: From 500 BC to the 1940s*, (Wiley-IEEE Press, 1999)
7. G.M. Masters, *Renewable and Efficient Electric Power Systems*, (Wiley, 2013)
8. M. Abdel-Salam, *High-Voltage Engineering: Theory and Practice, Revised and Expanded*, (CRC Press, 2000)
9. W. Berkson, *Fields of Force: The Development of A World View From Faraday to Einstein*, (Routledge, 1974) pp. 148
10. S. Jeszenszky, History of Transformers. IEEE Power Eng. Rev. **16**(12), 9 (1996)
11. M.H. Rashid, *Power Electronics Handbook: Devices, Circuits and Applications*, (Academic Press, 2010)
12. N. Mohan, T.M. Undeland, *Power Electronics: Converters, Applications, and Design*, (Wiley, 2007)
13. J.A. Baroudi, V. Dinavahi, A.M. Knight, A review of power converter topologies for wind generators. Renew. Energy **32**(14), 2369–2385 (2007)
14. P.W. Wheeler, J. Rodriguez, J.C. Clare, Matrix converters: a technology review. IEEE Trans. Ind. Electron. **49**(2), 276–288 (2002)
15. B. Zhang, D. Qiu, *Sneak Circuits of Power Electronic Converters*, (Wiley, 2014)
16. J.D. van Wyk, F.C. Lee, On A Future For Power Electronics. IEEE Trans. Emerg. Sel. Topics Power Electron. **1**(2), 59–72 (2013)
17. B.K. Bose, *Modern Power Electronics and AC Drives*, Vol. 123, (Prentice Hall, 2002)
18. P.R.A. Morris, *History of the World Semiconductor Industry*, (IET, 1990), pp. 11–25. ISBN 0863412270
19. T.L. Skvarenina, *The Power Electronics Handbook*, (CRC press, 2001)
20. B.K. Bose, *Power Electronics and Motor Drives: Advances and Trends*, (Academic Press, 2010)
21. P.K. Steimer, H.E. Gruning, J. Werninger, E. Carroll, S. Klaka, S. Linder, Igct-a new emerging technology for high-power, low-cost inverters. IEEE Ind. Appl. Mag. **5**(4), 12–18 (1999)
22. I. Colak, E. Kabalci, R. Bayindir, Review of multilevel voltage source inverter topologies and control schemes. Energy Convers. Manag. **52**(2), 1114–1128 (2011)
23. F.Z. Peng, Z-Source Inverter. IEEE Trans. Ind. Appl. **39**(2), 504–510 (2003)
24. Y. Siwakoti, F. Peng, and F. Blaabjerg, Impedance-source networks for electric power conversion part I: a topological review. IEEE Trans. Power Electron. **30**(2), 699–716 (2015)
25. Y. Siwakoti, F. Peng, F. Blaabjerg, Impedance-source networks for electric power conversion part II: review of control and modulation techniques. *IEEE Trans. Power Electron.* **30**(4), 1887–1906 (2015)
26. O. Ellabban, H. Abu-Rub, Z-source inverter: topology improvements review. IEEE Ind. Electron. Mag. **10**(1), 6–24 (2016)
27. Q.V. Tran, T.W. Chun, J.R. Ahn, H.H. Lee, algorithms for controlling both the DC boost and AC output voltage of Z-source inverter. IEEE Trans. Ind. Electron. **54**(5), 2745–2750 (2007)
28. U. Supatti, F.Z. Peng, Z-source inverter based wind power generation system in *Sustainable Energy Technologies 2008. ICSET 2008. IEEE International Conference*, (2008) pp. 634–638
29. R. Bharanikumar, R. Senthilkumar, A. Nirmal Kumar, Impedance source inverter for wind turbine driven permanent magnet generator, in *Power System Technology and IEEE Power India Conference, 2008. POWERCON 2008. Joint International Conference on*, (2008) pp. 1–7
30. F.Z. Peng, X. Yuan, X. Fang, Z. Qian, Z-source inverter for adjustable speed drives. IEEE Trans. Power Electron. Lett. **1**(2), 33–35 (2003)
31. F.Z. Peng, A. Joseph, J.Wang, M.S. Shen, L.H. Chen, Z.G Pan, E. Ortiz-Rivera, Y. Huang, Z-source inverter for motor drives. IEEE Trans. Power Electron. **20**(4), 857–863 (2005)

32. F.Z. Peng, A. Joseph, J.Wang, M. Shen, L. Chen, Z. Pan, E. Ortiz-Rivera, Y. Huang, Z-source inverter for motor drives. IEEE Trans. Power Electron. **20**(4), 857–863 (2005)
33. S.M. Dehghan, M. Mohamadian, A. Yazdian, F. Ashrafzadeh, A dual-input-dual-output Z-source inverter. IEEE Trans. Power Electron. **25**(2), 360–368 (2010)
34. S.M. Dehghan, M. Mohamadian, A. Yazdian, Hybrid electric vehicle based on bidirectional Z-source nine-switch inverter. IEEE Trans. Veh. Tech. **59**(6), 2641–2653 (2010)
35. F.Z. Peng, M. Shen, K. Holland, Application of Z-source inverter for traction drive of fuel cell battery hybrid electric vehicles. IEEE Trans. Power Electron. **22**(3), 1054–1061 (2007)
36. M.S. Shen, A. Joseph, J. Wang, F.Z. Peng, D.J. Adams, Comparison of traditional inverters and Z-source inverter for fuel cell vehicles. IEEE Trans. Power Electron. **22**(4), 1453–1463 (2007)
37. B.M. Ge, Q. Lei, W. Qian, F.Z. Peng, A family of Z-source matrix converters. IEEE Trans. Ind. Electron. **59**(1), 35–46 (2012)
38. M. Shen, A. Joseph, J. Wang, F.Z. Peng, D.J. Adams, Comparison of traditional inverters and Z-source inverter for fuel cell vehicles. IEEE Trans. Power Electron. **22**(4), 1453–1463 (2007)
39. Y. Huang, M. Shen, F.Z. Peng, J.Wang, Z-source inverter for residential photovoltaic systems. IEEE Trans. Power Electron. **21**(6), 1776–1782 (2006)
40. M. Hanif, M. Basu, K. Gaughan, Understanding the Operation of a Z-source inverter for photovoltaic application with a design example. IET Power Electron. **4**(3), 278–287 (2011)
41. F. Bradaschia, M.C. Cavalcanti, P.E.P. Ferraz, F.A.S. Neves, E.C. dos Santos, J.H.G.M. da Silva, Modulation for three-phase transformerless Z-source inverter to reduce leakage currents in photovoltaic systems. IEEE Trans. Ind. Electron. **58** (12), 5385–5395 (2011)
42. M.S. Shen, J. Wang, A. Joseph, F.Z. Peng, L.M. Tolbert, D.J. Adams, Constant boost control of the Z-source inverter to minimize current ripple and voltage stress. IEEE Trans. Ind. Appl. **42**(3), 770–778 (2006)
43. Y. Li, S. Jiang, J.G. Cintron-Rivera, F.Z. Peng, Modeling and control of quasi-Z-source inverter for distributed generation applications. IEEE Trans. Ind. Electron. **60**(4), 1532–1541 (2013)
44. B. Ge, H. Abu-Rub, F.Z. Peng, Q. Lei, A.T. de Almeida, F.J. Ferreira, D. Sun, Y. Liu, An energy-stored quasi-Z-source inverter for application to photovoltaic power system. IEEE Trans. Ind. Electron. **60**(10), 4468–4481 (2013)
45. J.C. Rosas-Caro, F.Z. Peng, H. Cha, C. Rogers, Z-source-converter-based energy-recycling zero-voltage electronic loads. IEEE Trans. Ind. Electron. **56**(12), 4894–4902 (2009)
46. Y. Li, J. Anderson, F.Z. Peng, D. Liu, Quasi-Z-source inverter for photovoltaic power generation systems, in *Applied Power Electronics Conference and Exposition, 2009. APEC 2009. Twenty-Fourth Annual IEEE*, vol. 4 no. 3 (2009) pp. 918–924
47. W.Y. Choi, C.G. Lee, Photovoltaic panel integrated power conditioning system using a high efficiency step-up DC-DC converter. Renew. Energy **41**, 227–234 (2012)
48. E.H. Ismail, M.A. Al-Saffar, A.J. Sabzali, A.A. Fardoun, A family of single-switch pwm converters with high step-up conversion ratio. IEEE Trans. Circ. Syst. I **55**(4), 1159–1171 (2008)

Part I
Voltage-, Current-, and Z-source Converters

In this part, preliminaries will be first given; and a profound analysis on voltage-, current- and Z-source converters will be carried out to understand the unique features of Z-source converters over that of voltage and current converters. Then follows a review on the currently popular Z-source converters.

Chapter 2
Voltage-, Current-, and Z-source Converters

Some fundamental concepts are to be introduced in this chapter, such as voltage sources, current sources, impedance networks, Z-source, two-port network, impedance source converters, impedance networks converters.

2.1 Voltage Sources and Current Sources

A power converter processes energy flow between two sources, i.e. generally between a generator and a load, as illustrated in Fig. 2.1. An ideal static converter is assumed to transmit electric energy between the two sources with 100% efficiency. The conversion efficiency is the main concern in designing a converter. Therefore, in practice, power converter design aims at improving the efficiency.

There are two types of sources, namely voltage and current sources, any of which could be either a generator or a load.

A real voltage source can be represented as an ideal voltage source in series with a resistance r_{VS}, with the ideal voltage source having zero resistance, to ensure its output voltage to be constant. The voltage source is normally equivalent to a capacitor C with infinite capacitance, i.e. $C = \infty$, so that $r_C = Z_C = -j\frac{1}{\omega C} \approx 0$, where Z_C denotes the resistance of the capacitor.

Similarly, a real current source can be represented as an ideal current source in parallel with a resistance r_{CS}, with the ideal current source having infinite resistance, so that its output current is constant, which is normally equivalent to an inductor with infinite inductance, i.e. $L = \infty$, which implies also $r_L = Z_L = j\omega L \approx \infty$, where Z_L represents the resistance of the inductor.

Correspondingly, converters are classified into voltage source converters and current source converters.

© Springer International Publishing AG 2018

G. Zhang et al., *Designing Impedance Networks Converters*, Studies in Systems, Decision and Control 119, DOI 10.1007/978-3-319-63655-9_2

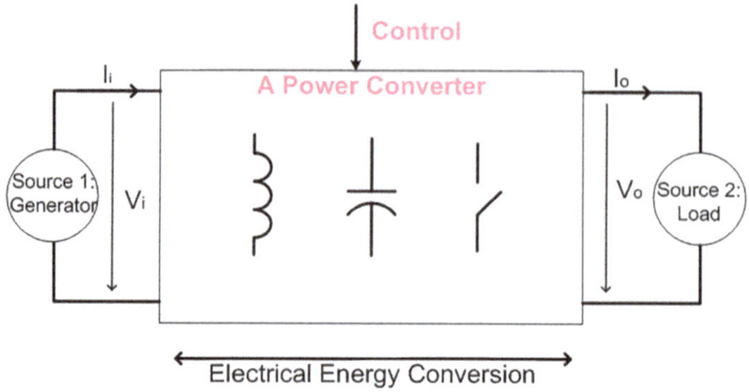

Fig. 2.1 A power converter

2.2 Impedance Network and Z-source

2.2.1 Impedance

The term, resistance, is just concerned with a DC circuit, which is extended to impedance in case of applying to both DC and AC circuits. Therefore, for DC circuits, resistance and impedance are equivalent. Unlike resistance, which has only magnitude and is represented as a *positive* real number (ohms (Ω)), impedance possesses both magnitude and phase and can be represented as a complex number with the imaginary part denoting the reactance and the real part representing the resistance.

Impedance is used to measure the opposition that a circuit presents to a current when a voltage is applied [1], and is defined as the frequency domain ration of the voltage to the current. For a sinusoidal current or voltage input, the polar form of the complex impedance relates the amplitude and phase of the voltage and current. In particular,

- the magnitude of the complex impedance is the ratio of the voltage amplitude to the current amplitude, and
- the phase of the complex impedance is the phase shift by which the current lags or leads the voltage.

2.2.2 Impedance Network and Two-Port Network

Like a resistor network, which is a collection of interconnected resistors in series or/and parallel, an impedance network in the context of power electronics, which involves nonlinear switches, is a network of impedance components like switches, sources, inductors, and capacitors, interconnected in series or/and parallel. An

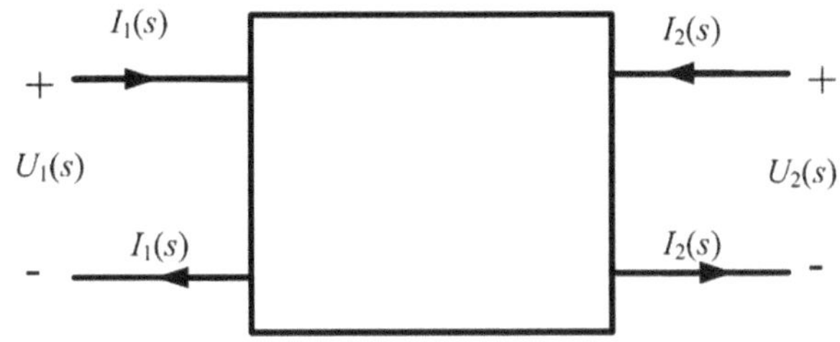

Fig. 2.2 Two-port network

impedance network can be passive, if it is just composed of inductors and/or capacitors, or active, if it is constituted of switches and/or diodes, inductors and/or capacitors.

It is difficult, if not impossible, to analyse an impedance network using (linear) circuit theory due to the nonlinear switching components in the impedance network. It is, however, helpful to simplify the analysis of an impedance network by reducing the number of its components, which is then normally done by replacing the actual components with notional components of the same functions. Among existing analysis methods, such as Nodal and Mesh methods [2], the two-port network is well suited for the analysis of the impedance network [3].

A two-port network, as shown in Fig. 2.2, is an electrical network or a device with four terminals, which are arranged into two pairs called ports, i.e. each pair of terminals is one port. As shown in Fig. 2.2, the left port is usually regarded as the input port, while the right one is the output port. Therefore, a two-port network is represented by four external variables, i.e. voltage $U_1(s)$ and current $I_1(s)$ at the input port, and voltage $U_2(s)$ and current $I_2(s)$ at the output port, so that the two-port network can be treated as a black box modeled by the relationships between the four variables $U_1(s)$, $I_1(s)$, $U_2(s)$, and $I_2(s)$ [4–7].

The transmission equation of a two-port network is given by [8–11]

$$\begin{bmatrix} U_1(s) \\ I_1(s) \end{bmatrix} = \mathbf{A}(s) \cdot \begin{bmatrix} U_2(s) \\ -I_2(s) \end{bmatrix}, \tag{2.1}$$

where $\mathbf{A}(s)$ is the transmission matrix and written as

$$\mathbf{A}(s) = \begin{bmatrix} A_{11}(s) & A_{12}(s) \\ A_{21}(s) & A_{22}(s) \end{bmatrix}, \tag{2.2}$$

whose elements are defined as

$$\begin{cases} A_{11}(s) = \left.\dfrac{U_1(s)}{U_2(s)}\right|_{I_2(s)=0} , \\ A_{12}(s) = \left.\dfrac{U_1(s)}{-I_2(s)}\right|_{U_2(s)=0} , \\ A_{21}(s) = \left.\dfrac{I_1(s)}{U_2}\right|_{I_2(s)=0} , \\ A_{22}(s) = \left.\dfrac{I_1(s)}{-I_2(s)}\right|_{U_2(s)=0} . \end{cases} \tag{2.3}$$

Therefore, (2.1) can be rewritten as

$$\begin{cases} U_1(s) = A_{11}(s)U_2(s) + A_{12}(s)(-I_2(s)) , \\ I_1(s) = A_{21}(s)U_2(s) + A_{22}(s)(-I_2(s)) . \end{cases} \tag{2.4}$$

A two-port network model is a mathematical circuit analysis technique to represent a complex circuit into a simple notation. A two-port network is regarded as a "black box" with its properties specified by a matrix of numbers, which allows the response of the network to signals applied to the ports to be calculated easily, without solving all the internal voltages and currents in the network [7].

Impedance networks can have multiple ports connecting external circuits, but generally have two ports, and can thus be equivalent to a two-port network. In terms of Thevenin's equivalent impedance theorem, the *input impedance* of a two-port network is the equivalent impedance of the two-port network with an open input port and an output port connecting a load; while the *output impedance* (also named as *source impedance* or *internal impedance*) is the equivalent impedance of the two-port network with a short-circuited input port and an open output port. Further in terms of Ohm's law, the input impedance of a two-port network $Z_i(s)$ reads

$$Z_i(s) = \frac{U_1(s)}{I_1(s)} = \frac{A_{11}(s)Z_L(s) + A_{12}(s)}{A_{21}(s)Z_L(s) + A_{22}(s)} , \tag{2.5}$$

where $Z_L(s)$ is the load impedance of the two-port network's output port.

Similarly, the output impedance of two-port network $Z_o(s)$ writes

$$Z_o(s) = \frac{U_2(s)}{I_2(s)} = \frac{A_{22}(s)Z_S(s) + A_{12}(s)}{A_{21}(s)Z_S(s) + A_{11}(s)} , \tag{2.6}$$

where $Z_S(s)$ is the source impedance of the two-port network's input port.

2.2.3 *Impedance Source Converters (Z-source Converters)*

An impedance network together with a source constitute an *impedance source* (also named a Z-source), with its equivalent impedance $Z \in [0, +\infty)$. The impedance source is a general source in the sense that it includes voltage- and current sources as its extreme cases; that is, it becomes a voltage source for $Z = 0$; and a current source for $Z \rightarrow \infty$. It can then exhibit rich properties for $0 \leq Z < \infty$.

Correspondingly, an impedance source converter is thus coined, which possesses unique advantages over traditional voltage and current sources converters and can well meet more stringent requirements from today's industry. It is known that voltage source converters suffer from shoot-through problems, the inapplicability to a capacitive load, and limited gains of output voltages; while current source converters have open-circuit problems, the inapplicability to an inductive load, and limited gains of output currents. A well-designed impedance source converter can overcome those above-mentioned problems.

2.2.4 *Impedance Networks Converters*

Extending from impedance source converters, an impedance network can be put in other positions of the converter but not only with the source, and these kinds of converters are named as impedance network converter. The impedance networks converter is a general one in the sense that it include impedance source converters when the impedance network is placed with a source.

2.3 Voltage-Source- and Current-Source-Inverters

A converter is a general term for AC-DC rectifiers, DC-DC choppers, DC-AC inverters, and AC-AC converters. AC-DC rectifiers and AC-AC converters may have the problems of shoot-through, open-circuit and limited output gains; while DC-DC choppers may suffer from the shoot-through and open-circuit problems and inapplicability to a capacitive or inductive load, as well as DC-AC inverters may have all of the above-mentioned problems. For simplicity, voltage-source- and current-source-inverters are taken as examples to be qualitatively analysed from the perspective of impedance networks.

Voltage-source- and current-source-inverters are depicted in Fig. 2.3, where $V_{\mathrm{VS}}(s)$ and $I_{\mathrm{VS}}(s)$ in Fig. 2.3a represent the voltage and current of the voltage source; while $V_{\mathrm{CS}}(s)$ and $I_{\mathrm{CS}}(s)$ in Fig. 2.3b stand for the voltage and current of the current source, respectively. Furthermore, their equivalent circuits are drawn in Fig. 2.4, where $Z_{\mathrm{VS}}(s)$ and $Z_{\mathrm{L}}(s)$ are the equivalent source impedance and equivalent load impedance of the voltage source inverter in Fig. 2.4a, whose corresponding two-port

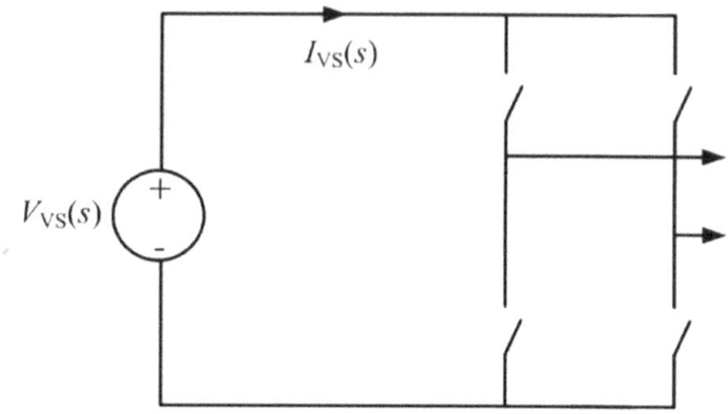

(a) Voltage source inverters

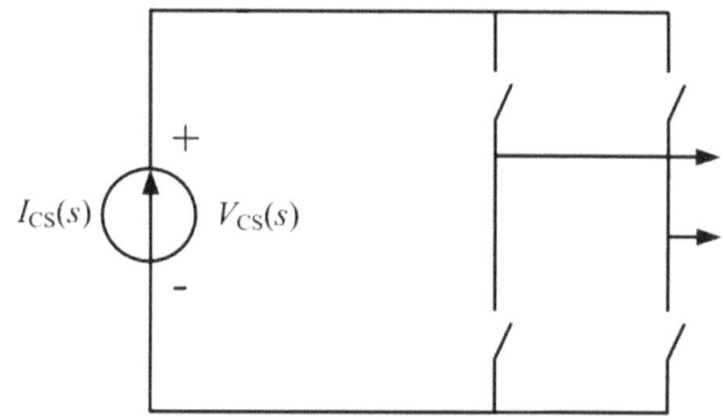

(b) Current source inverters

Fig. 2.3 Voltage source and current source inverters

network is indicated in the dashed box in Fig. 2.4a, where $Z_{VS}(s)$ is the unique component in the two-port network; while $Y_{CS}(s)$ and $Y_L(s)$ are the equivalent source admittance and load admittance of the current source inverter in Fig. 2.4b, whose corresponding two-port network is shown in the dashed box in Fig. 2.4b, where $Y_{CS}(s)$ is also the unique component in the two-port network.

2.4 Voltage Source Inverters

2.4.1 Shoot-Through

In terms of (2.3), the transmission matrix of the voltage source inverter in Fig. 2.4a reads

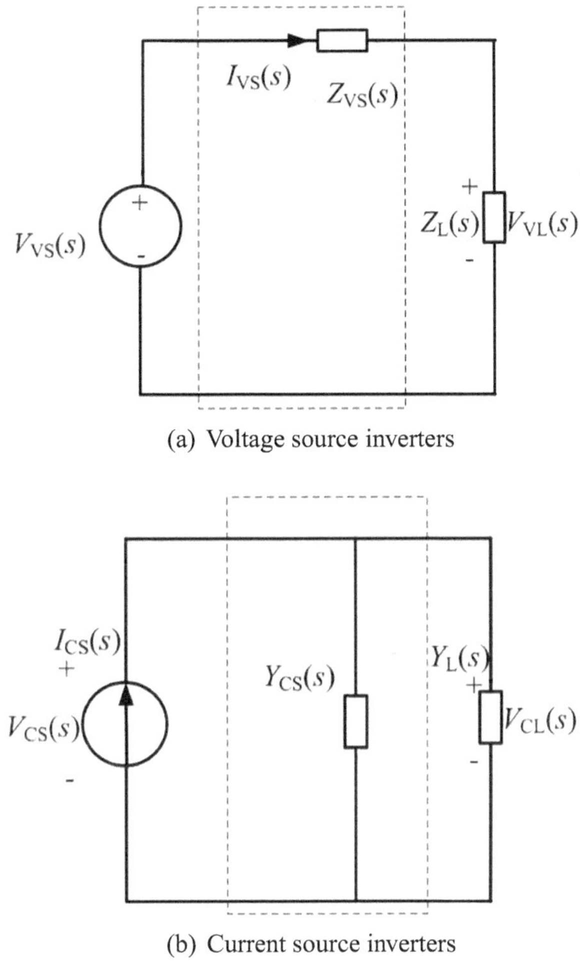

Fig. 2.4 Equivalent circuits of voltage-source- and current-source-inverters with two-port networks

$$\begin{cases} A_{V11}(s) = 1 \, , \\ A_{V12}(s) = Z_{VS}(s) \, , \\ A_{V21}(s) = 0 \, , \\ A_{V22}(s) = 1 \, . \end{cases} \tag{2.7}$$

Substituting (2.7) into (2.5) results in the input impedance of the voltage source inverter as

$$Z_i(s) = \frac{A_{V11}(s)Z_L(s) + A_{V12}(s)}{A_{V21}(s)Z_L(s) + A_{V22}(s)} = Z_L(s) + Z_{VS}(s) \, , \tag{2.8}$$

while the input current of the voltage source is thus obtained as

$$I_{VS}(s) = \frac{V_{VS}(s)}{Z_i(s)} = \frac{V_{VS}(s)}{Z_L(s) + Z_{VS}(s)} . \tag{2.9}$$

It is obvious that $Z_L(s) = 0$ in case that the switches of the voltage source inverter on a bridge are turned on simultaneously. Moreover, the source impedance $Z_{VS}(s)$ is normally very small, i.e. $Z_{VS}(s) \approx 0$. Therefore, $Z_i(s) = Z_L(s) + Z_{VS}(s) \approx 0$, which implies $I_{VS}(s) \to \infty$. Thus, the voltage source is shorted and a very large current will break down the switches. This is the so-called **shoot-through problem**.

In order to prevent the occurrence of shoot-through, the dead-time compensation technique has often been adopted to prevent switches from turning on simultaneously [12].

2.4.2 *Limited Output Voltage Gain*

In terms of Fig. 2.4a, substituting $Z_S(s) = 0$ and (2.7) into (2.6) results in its output impedance as

$$Z_o(s) = \frac{A_{V22}(s)Z_S(s) + A_{V12}(s)}{A_{V21}(s)Z_S(s) + A_{V11}(s)} = Z_{VS}(s) . \tag{2.10}$$

Obviously, the voltage of the load can be expressed as

$$V_{VL}(s) = V_{VS}(s) - I_L(s)Z_{VS}(s) . \tag{2.11}$$

It is straightforward from (2.11) that $Z_{VL}(s) \leq V_{VS}(s)$ due to $Z_{VS}(s) \geq 0$ and $I_L(s) \geq 0$; namely, the load voltage $V_{VL}(s)$ is lower than or equal to the source voltage $V_{VS}(s)$.

In order to fulfill the high output voltage gain requirements in industrial applications like solar energy applications, DC-DC boost front stage converters can be cascaded to boost the output voltage, which has actually changed its output impedance features [13–16].

2.4.3 *Inapplicability to Capacitive Loads*

It is known that the electrical loads can be classified into resistive, capacitive, and inductive ones. A capacitive load is an AC electrical load, in which the current reaches its peak before the voltage; while an inductive load is a load that pulls a large amount of current when first energised, for example, motors, transformers, and wound control gear, and a resistive load is a load which consumes electrical energy in a sinusoidal manner. This means that the current flow is in time with and directly proportional to the voltage, such as incandescent lighting and electrical heaters.

The impedance $Z_{\mathrm{VS}}(s)$ in a two-port network is equivalent to a capacitor with very large capacitance, which implies that $Z_{\mathrm{VS}}(s) = -j\frac{1}{\omega C} \approx 0$. In term of (2.11), one has $V_{\mathrm{VL}}(s) = V_{\mathrm{VS}}(s)$. It is remarked if the load impedance $Z_{\mathrm{L}}(s)$ is capacitive, a capacitive source offers energy to a capacitive load, while $V_{\mathrm{VL}}(s) = V_{\mathrm{VS}}(s)$ at a steady state implies that the voltage source inverter does not function, and is thus inapplicable to capacitive loads.

It is concluded that, due to the impedance of a two-port network between the voltage source and the inverter bridges, the voltage source inverter has the problems of the shoot-through, limited output voltage gains, and inapplicability to capacitive loads, which restrain its wide applications.

2.5 Current Source Inverters

2.5.1 Open-Circuit

In terms of (2.3), the transmission matrix of the current source inverter in Fig. 2.4b reads

$$\begin{cases} A_{\mathrm{C11}}(s) = 1\,, \\ A_{\mathrm{C12}}(s) = 0\,, \\ A_{\mathrm{C21}}(s) = Y_{\mathrm{CS}}(s)\,, \\ A_{\mathrm{C22}}(s) = 1\,, \end{cases} \tag{2.12}$$

where $Y_{\mathrm{CS}}(s)$ is the source admittance of the current source inverter, which is reciprocal to its source impedance.

Substituting (2.12) into (2.5) results in the input admittance of the current source inverter

$$Y_{\mathrm{i}}(s) = \frac{1}{Z_{\mathrm{i}}(s)} = \frac{A_{\mathrm{C21}}(s)\dfrac{1}{Y_{\mathrm{L}}(s)} + A_{\mathrm{C22}}(s)}{A_{\mathrm{C11}}(s)\dfrac{1}{Y_{\mathrm{L}}(s)} + A_{\mathrm{C12}}(s)} = Y_{\mathrm{L}}(s) + Y_{\mathrm{CS}}(s)\,, \tag{2.13}$$

where $Y_{\mathrm{L}}(s)$ and $Y_{\mathrm{CS}}(s)$ are the load and source admittances, respectively, as shown in Fig. 2.4b, while the input voltage of the current source is thus obtained as

$$V_{\mathrm{CS}}(s) = \frac{I_{\mathrm{CS}}(s)}{Y_{\mathrm{L}}(s) + Y_{\mathrm{CS}}(s)}\,, \tag{2.14}$$

where $I_{CS}(s)$ is the current of current source, as shown in Fig. 2.4b.

An inverter normally includes at least one inverter bridge, while one inverter bridge is normally composed of one upper switch and one lower switch. On each bridge, either the upper switch or the lower switch must be kept on; otherwise, one has $Y_L(s) = 0$. Moreover, the source admittance $Y_{CS}(s)$ is normally very small, i.e. $Y_{CS}(s) \approx 0$. Therefore, $Y_i(s) = Y_L(s) + Y_{CS}(s) \approx 0$, which implies $V_{CS}(s) \to \infty$. Thus, the current source is open-circuit and a very large voltage will break down the switches.

In order to prevent the open-circuit problem, the overlapped time technique on upper and lower switches has been normally utilized to ensure at least one of the upper switches and one of the lower switches to be on at any time [12].

2.5.2 Limited Output Current Gain

In terms of (2.6), one can obtain the output admittance of the current source inverter as

$$Y_o(s) = \frac{1}{Z_o(s)} = \frac{A_{C21}(s)\dfrac{1}{Y_{CS}(s)} + A_{C11}(s)}{A_{C22}(s)\dfrac{1}{Y_{CS}(s)} + A_{C12}(s)} = Y_{CS}(s)\,, \tag{2.15}$$

while the output current is

$$I_{CL}(s) = I_{CS}(s) - V_{CS}(s)Y_{CS}(s)\,. \tag{2.16}$$

For $V_{CS}(s) \geq 0$ and $Y_{CS} \geq 0$, one has $I_{CL} \leq I_{CS}$, namely, the load current $I_{CL}(s)$ is lower than or equal to the source current $I_{CS}(s)$.

2.5.3 Inapplicability to Inductive Loads

The admittance $Y_{CS}(s)$ in a two-port network is equivalent to an inductor with very large inductance, which implies that $Y_{CS}(s) = -j\frac{1}{\omega L} \approx 0$. It is remarked if the load admittance $Y_L(s)$ is inductive, an inductive source offers energy to an inductive load, while $I_{CL}(s) = I_{CS}(s)$ at a steady state implies that the current source inverter does not work and is thus inapplicable to inductive loads.

It is concluded that, due to the admittance of the two-port network between the current source and the inverter bridges, the current source inverter has the problems of open-circuit, limited output current gains, and inapplicability to inductive loads.

2.6 Z-source Inverters

Peng [17] has proposed to use an impedance network (named as Z-network) in 2002, as shown in Fig. 2.6, to couple with a DC source to form a novel source, as shown in the rectangles in Fig. 2.7, including voltage- and current-type Z-source inverters. Applying this Z-source technology in other converters results in Z-source DC-DC converters (Fig. 2.5a), Z-source AC-DC rectifiers (Fig. 2.5b), and Z-source AC-AC converters (Fig. 2.5c).

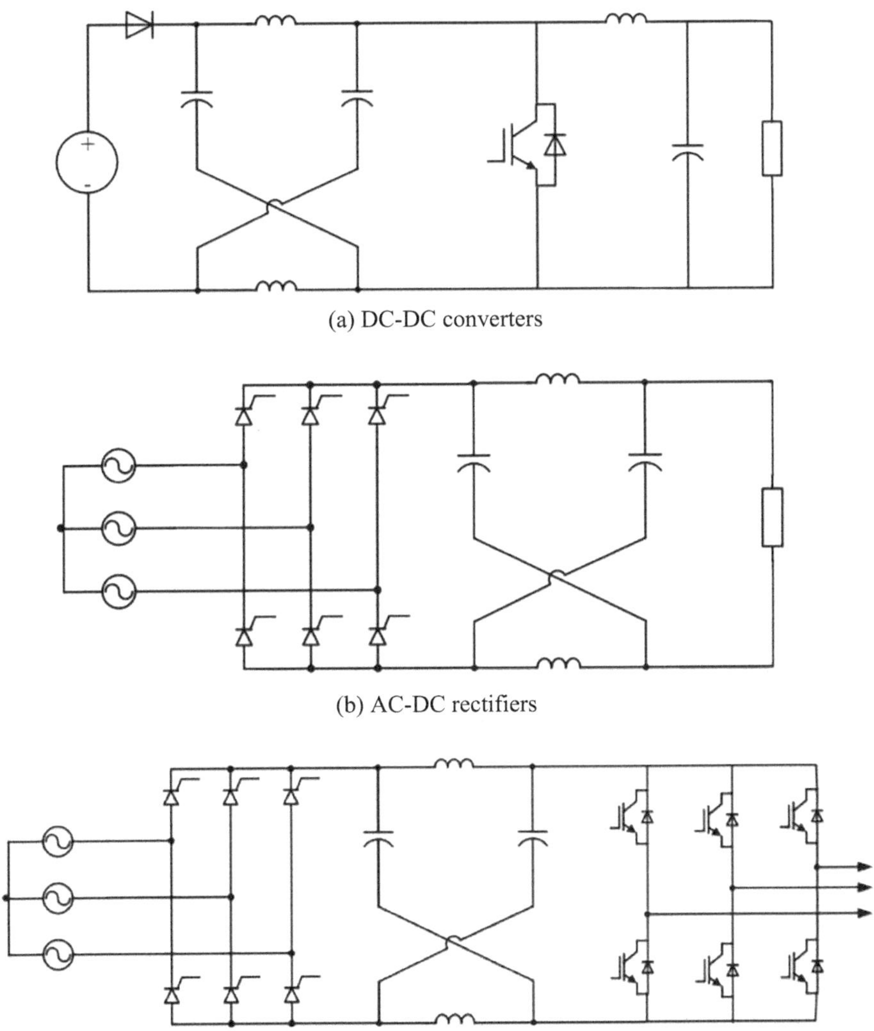

(a) DC-DC converters

(b) AC-DC rectifiers

(c) AC-AC converters

Fig. 2.5 Other typical Z-source converters

Fig. 2.6 A Z-network

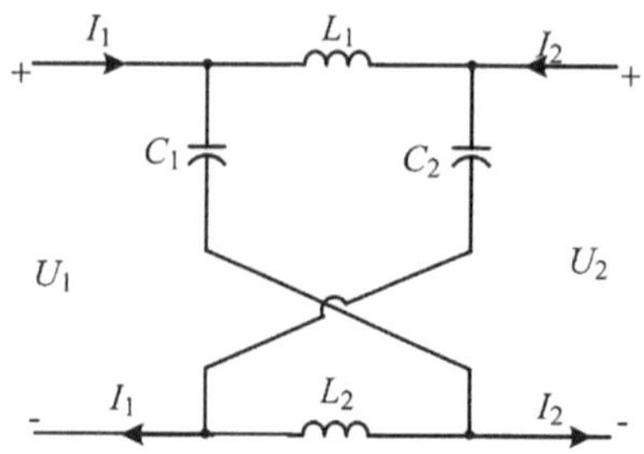

Fig. 2.7 Z-source inverters

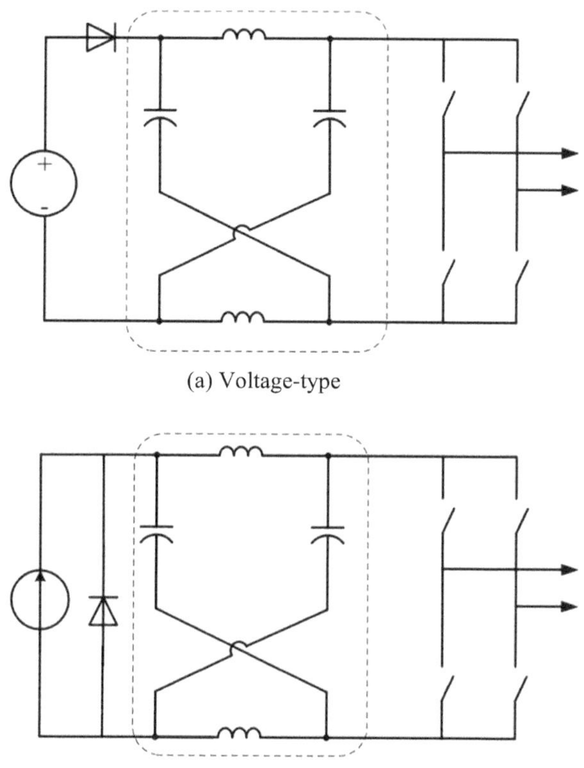

(a) Voltage-type

(b) Current-type

Similarly, voltage-type Z-source inverters are also taken as examples, for simplicity, to explain the reasons that Z-source converters can overcome the problems of voltage source and current source converters. The diagram of a voltage-type Z-source inverter is drawn in Fig. 2.7a, whose equivalent two-port network is illustrated in the dashed box in Fig. 2.8.

Assume $L_1 = L_2 = L$ and $C_1 = C_2 = C$, and denote the impedance of diode D by $Z_{ZS}(s)$.

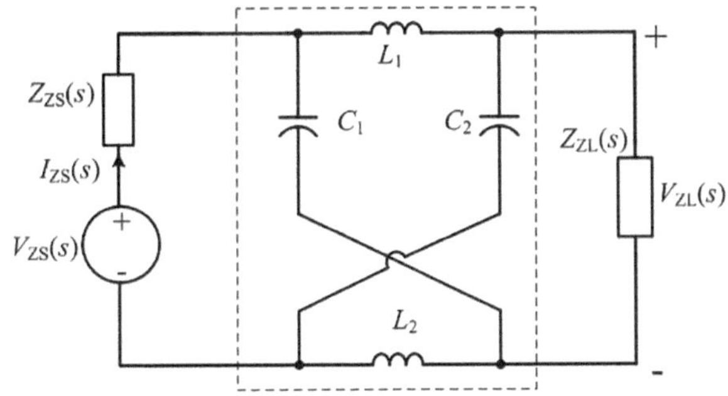

Fig. 2.8 Equivalent circuit of voltage-type Z-source inverters with two-port network

In terms of (2.2), one can obtain the transmission matrix of the Z-network as follows

$$\mathbf{A}_Z(s) = \begin{bmatrix} A_{Z11}(s) & A_{Z12}(s) \\ A_{Z21}(s) & A_{Z22}(s) \end{bmatrix}, \tag{2.17}$$

where, in terms of (2.3), the elements write

$$\begin{cases} A_{Z11}(s) = \dfrac{1+s^2LC}{1-s^2LC}, \\ A_{Z12}(s) = \dfrac{2sL}{1-s^2LC}, \\ A_{Z21}(s) = \dfrac{2sC}{1-s^2LC}, \\ A_{Z22}(s) = \dfrac{1+s^2LC}{1-s^2LC}. \end{cases} \tag{2.18}$$

Substituting $Z_S(s) = Z_{ZS}(s)$, $Z_L(s) = Z_{ZL}(s)$ and (2.18) into (2.5) and (2.6) results in the input and output impedances of the Z-network as

$$\begin{cases} Z_{Zi}(s) = \dfrac{A_{Z11}(s)Z_{ZL}(s) + A_{Z12}(s)}{A_{Z21}(s)Z_{ZL}(s) + A_{Z22}(s)} = \dfrac{(s^2LC+1)Z_{ZL}(s) + 2sL}{s^2LC + 2sCZ_{ZL}(s) + 1}, \\ Z_{Zo}(s) = \dfrac{A_{Z22}(s)Z_{ZS}(s) + A_{Z12}(s)}{A_{Z21}(s)Z_{ZS}(s) + A_{Z11}(s)} = \dfrac{(s^2LC+1)Z_{ZS}(s) + 2sL}{s^2LC + 2sCZ_{ZS}(s) + 1}, \end{cases} \tag{2.19}$$

where $Z_{ZS}(s)$ is the source impedance of the input port of the Z-network and $Z_{ZL}(s)$ is the load impedance of the output port of the Z-network, described as

$$Z_{ZS}(s) = \begin{cases} 0, & \text{if } D \text{ is on}, \\ \infty, & \text{otherwise}, \end{cases} \tag{2.20}$$

and

$$Z_{\rm ZL}(s) = \begin{cases} 0, & \text{at a shoot-through state,} \\ \infty, & \text{at an open-circuit state,} \\ Z_{\rm Z}(s), & \text{at a normal state,} \end{cases} \tag{2.21}$$

where $Z_{\rm Z}(s)$ is the load impedance of the inverter bridge.

Substituting (2.20) and (2.21) into (2.19) leads to the input and output impedances as

$$Z_{\rm Zi}(s) = \begin{cases} \dfrac{2sL}{s^2LC+1}, & \text{at a shoot-through state,} \\ \dfrac{s^2LC+1}{2sC}, & \text{at an open-circuit state,} \\ \dfrac{(s^2LC+1)Z_{\rm Z}(s)+2sL}{2sCZ_{\rm Z}(s)+s^2LC+1}, & \text{at a normal state,} \end{cases} \tag{2.22}$$

and

$$Z_{\rm Zo}(s) = \begin{cases} \dfrac{2sL}{s^2LC+1}, & \text{if } D \text{ is on,} \\ \dfrac{s^2LC+1}{2sC}, & \text{otherwise.} \end{cases} \tag{2.23}$$

2.6.1 *Immunity to the Shoot-Through*

The input current of the Z-source inverter is expressed as

$$I_{\rm ZS}(s) = \frac{V_{\rm ZS}(s)}{Z_{\rm Zi}(s)}, \tag{2.24}$$

where $Z_{\rm ZL}(s) = 0$ if the switches on a bridge are turned on simultaneously. It is obvious that $Z_{\rm Zi}(s) \neq 0$ holds in all cases in terms of (2.22). Therefore, the Z-source inverter can operate at shoot-through states. Compared to the voltage source inverter, Z-source inverter is immune to the shoot-through problem, so that the short-circuited phenomenon at the source can be avoided because the Z-network increases the input impedance.

2.6.2 *High Output Voltage Gains*

Denote the duty cycle of the diode D as d and assume $d \in [0, 1]$. In terms of (2.23), one can obtain the average output impedance as

$$Z_{Zo}(s) = \frac{(1-d)L}{2}\left(\frac{s^4 + s^2\left(\dfrac{2(1+d)}{(1-d)LC}\right) + \dfrac{1}{L^2C^2}}{s^3 + s\dfrac{1}{LC}}\right), \tag{2.25}$$

while the output voltage of the Z-source inverter, $V_{ZL}(s)$, is expressed as

$$V_{ZL}(s) = V_{ZS}(s) - I_{ZL}(s)Z_{Zo}(s). \tag{2.26}$$

It is obvious that $Z_{Zo}(s)$ is the function of the duty d in terms of (2.25). Adjusting $Z_{Zo}(s)$ to be negative or positive via d, one can obtain either $V_{ZL}(s) > V_{ZS}(s)$ or $V_{ZL}(s) < V_{ZS}(s)$, which implies that the Z-source inverters can overcome the limited voltage gains of traditional voltage source inverters.

2.6.3 *Applicability both to Capacitive and Inductive Loads*

Assume that $Z_Z(s)$ is capacitive. Then, in terms of (2.23), one has

$$Z_Z(s) = \frac{1}{sC_L}, \tag{2.27}$$

where C_L is the capacitance of the load.

By adjusting the duty d, and the inductance L, capacitance C of the Z-network, the output impedance of the Z-network can exhibit the inductive feature, implying that the Z-source inverter is applicability of a capacitive load.

Similarly, assume that $Z_Z(s)$ is inductive and one can also prove that the Z-source inverter is also applicability of an inductive load.

It is thus concluded that due to the embedded Z-network, Z-source inverters have unique advantages over traditional ones, i.e. immunity to the shoot-through, higher output voltage gains, and applicability of both capacitive and inductive loads, which have a great potential in renewable energy applications.

References

1. Wikipedia, Electrical Impedance, www.en.wikipedia.org/wiki/Electrical_impedance
2. T.B.M. Neill, Generalisation of nodal and mesh analysis. Electron. Lett. Vol. **5**(16), 365–366
3. S.S. Haykin, Active network theory (Addison-Wesley, 1970)
4. W.F. Egan, Practical RF system design (Wiley-IEEE 2003)
5. P.R. Gray, P.J. Hurst, S.H. Lewis, R.G. Meyer, *Analysis and Design of Analog Integrated Circuits. 4th* (Wiley, New York, 2001)

6. R.C. Jaeger, T.N. Blalock, *Microelectronic Circuit Design*, 3rd edn. (Boston: McGraw-Hill Press, 2006)
7. Wikipedia, Two-port network, http://en.wikipedia.org/wiki/Two-port_network
8. Y.J. Matthaei, *Microwave Filters, Impedance-Matching Networks, and Coupling Structures* (McGraw-Hill Press, New York, 1964)
9. S. Ghosh, *Network Theory: Analysis and Synthesis* (Prentice Hall of India, India, 2005)
10. P.S. Farago, *An Introduction to Linear Network Analysis*. (The English Universities Press Ltd, 1961)
11. H.J. Carlin, P.P. Civalleri, *Wideband Circuit Design*, (CRC Press, 1998)
12. L. Chen, F.Z. Peng, Dead-Time elimination for voltage source inverters. IEEE Trans. Power Electron. **23**(2), 574–580 (Feb. 2008)
13. P.W. Sun, Cascade dual-buck inverters for renewable energy and distributed generation. Ph.D. Dissertation, Virginia Polytechnic Institute and State University, 2012
14. L. Wang, Study of the cascaded Z-source inverter to solve the partial shading for the grid-connected PV system. M.Sc. Dissertation, Florida State University, 2010
15. D. Persson, Islanding detection in power electronic converter based distributed generation. M.Sc. Dissertation, Lund University, 2007
16. C. Pekuz, Z-source full-bridge DC-DC converter, M. Sc. Dissertation, Middle East Technical University 2010
17. F.Z. Peng, Z-Source Inverter, IEEE Trans. Ind. Appl. **39**(2), 504–510, (Mar. 2003)

Chapter 3
Impedance Source Converters: State-of-the-Art

This chapter presents the-stat-of-the-art of impedance source converters to understand their respective features and application scenarios [1, 2].

3.1 Quasi-Z-source Converters

The quasi-Z-source converter was proposed by Anderson and Peng in 2008 for applications of motor systems, new energy systems, and micro-grid systems. According to the operational modes in voltage-type or current-type and continuous or discontinuous current, quasi-Z-source converters can be classified into four categories, i.e. voltage-fed quasi-Z-source inverters with continuous input current, voltage-fed quasi-Z-source inverters with discontinuous input current, current-fed quasi-Z-source inverters with continuous input current, and current-fed quasi-Z-source inverters with discontinuous input current, which are shown in Fig. 3.1 [3, 4]. It is found by Cao and Peng [5] that all of the impedance (Fig. 3.2) networks in Fig. 3.1 can be derived from the one in Fig. 2.6. For instance, a voltage-fed quasi-Z-source inverter with continuous input current in Fig. 3.1a is equivalent to that in Fig. 3.3, whose switches S_1 and S_2 are equivalent to the diode D and the inverting bridge in Fig. 3.1b, respectively.

It is remarked that the impedance network in Fig. 3.2 is a typical quasi-Z-network, based on which various quasi-Z-networks can be derived. For example, Cao and Peng have proposed a family of quasi-Z-source DC-DC converters [5–11], and Vinikov et.al. have also proposed some novel quasi-Z-source DC-DC converters for renewable energy systems [12–16].

© Springer International Publishing AG 2018

G. Zhang et al., *Designing Impedance Networks Converters*, Studies in Systems, Decision and Control 119, DOI 10.1007/978-3-319-63655-9_3

Fig. 3.1 Quasi-Z-source inverters [3]

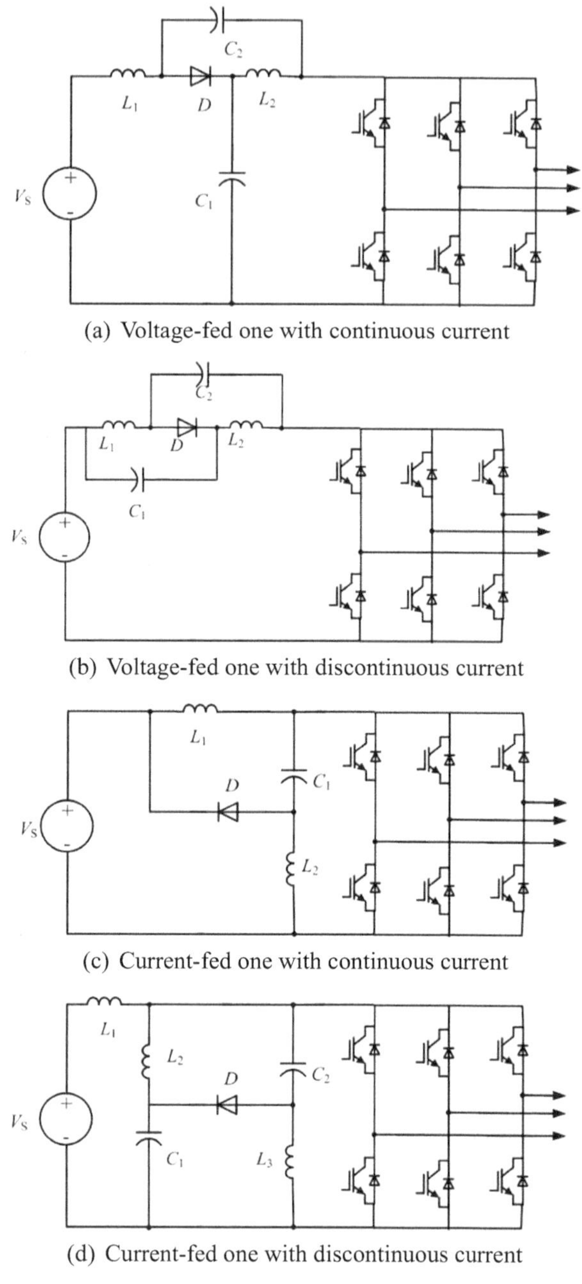

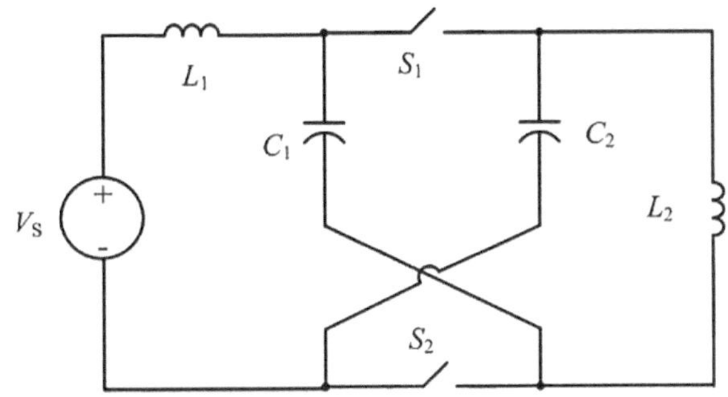

Fig. 3.2 Equivalent circuit of the converter in Fig. 3.1 [5]

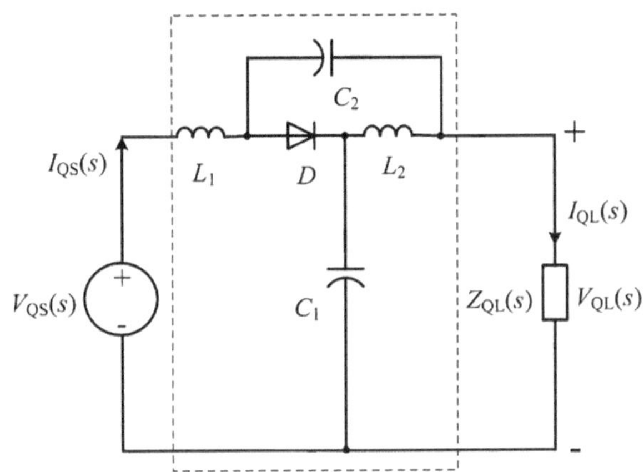

Fig. 3.3 Equivalent circuit of voltage-fed quasi-Z-source inverters with two-port network

3.2 Trans-Z-source Converters

Compared with traditional voltage source converters, whose voltage gains are normally in scale of 5 $\sim$ 6, typical Z-source and quasi-Z-source converters can reach much larger voltage gains in scale of 20, which are, however, still not large enough for some special applications. For example, voltage gains of converters utilized in solar energy systems need to reach the scales of decades or even hundreds.

In 2010, Qian and Peng et al. have integrated the transformers or coupled inductors into the impedance networks (shown in Fig. 2.6) and quasi-impedance-networks (shown in Fig. 3.2) to construct trans-impedance-networks as shown in the dashed box in Fig. 3.4 [17], and thus various trans-Z-source converters can be coined.

In terms of different operational modes of input current and coupled inductors, trans-Z-source converters can be classified into six categories, i.e. voltage-fed trans-quasi-Z-source inverters, current-fed trans-quasi-Z-source inverters, voltage-fed trans-quasi-Z-source inverters with coupled inductors, current-fed trans-quasi-Z-source inverters with two coupled inductors, voltage-fed trans-quasi-Z-source inverters, and current-fed trans-quasi-Z-source inverters, as shown in Fig. 3.4. Therein, trans-Z-source converters not only maintain the main features of traditional Z-source

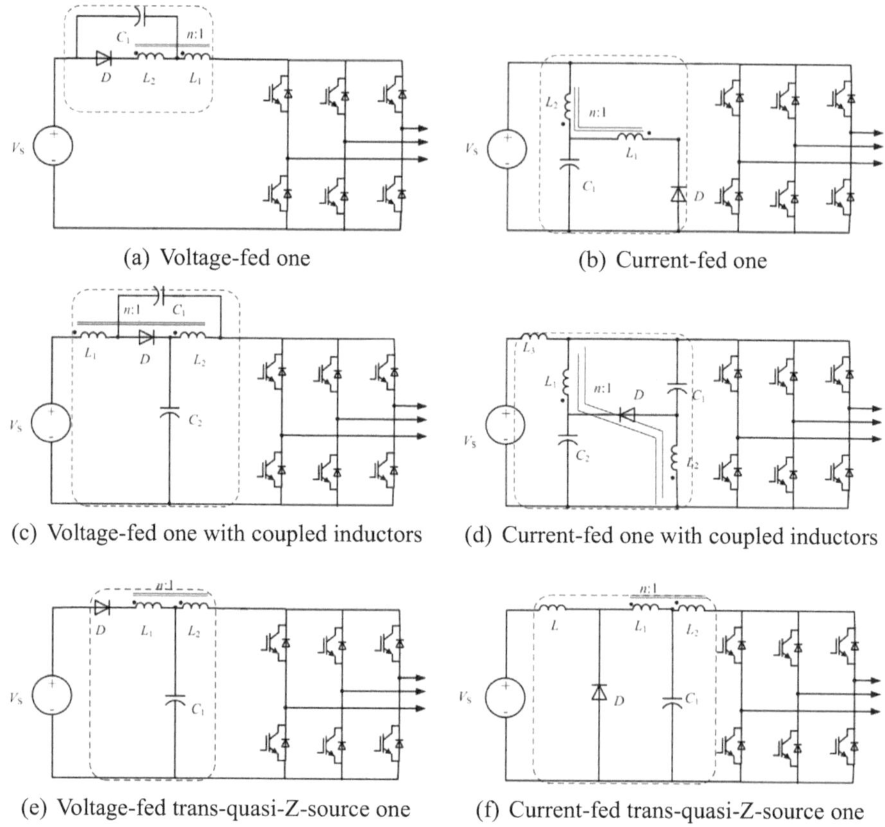

(a) Voltage-fed one (b) Current-fed one

(c) Voltage-fed one with coupled inductors (d) Current-fed one with coupled inductors

(e) Voltage-fed trans-quasi-Z-source one (f) Current-fed trans-quasi-Z-source one

Fig. 3.4 Typical trans-Z-source converters [17]

converters, but also exhibit some unique advantages, i.e. increased voltage gains and reduced voltage stress in the voltage-fed trans-Z-source inverters due to the transformers or coupled inductors, and the expanded operation quadrant (Fig. 3.5) in the current-fed trans-Z-source inverters. However, transformers and coupled inductors increase volume and cost.

3.3 Embedded-Z-source Converters

In order to obtain smaller volume and higher robustness, P.C. Loh et al. proposed embedded-Z-source converters in 2010 [18]. Instead of using an external LC filter, they proposed an alternative family of embedded-Z-source inverters, which adopts the concept of embedding the input DC sources into the LC impedance network, using its existing inductive elements for current filtering in voltage-type embedded-

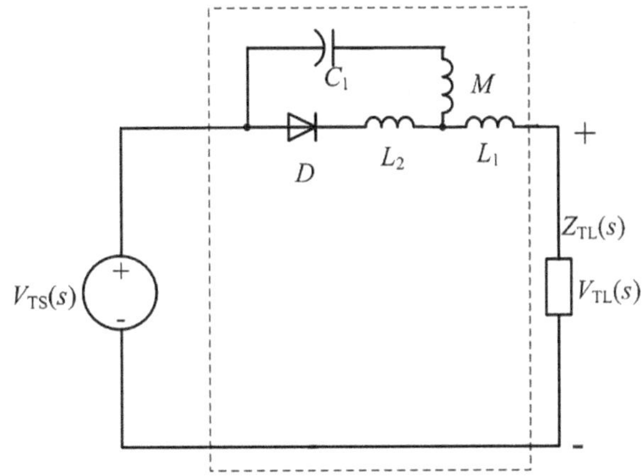

Fig. 3.5 Equivalent circuit of trans-Z-source inverters with two-port networks

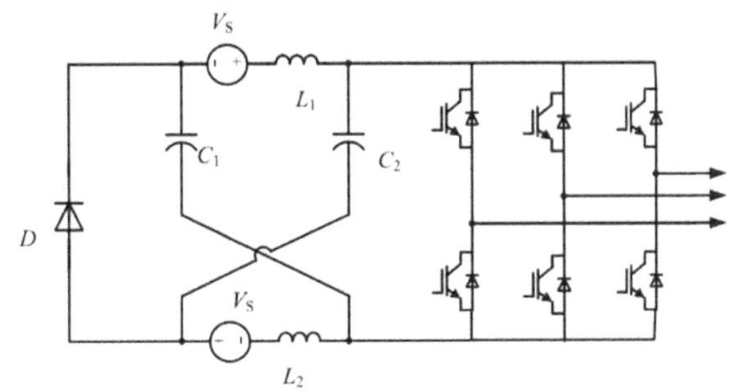

(a) Two-level type

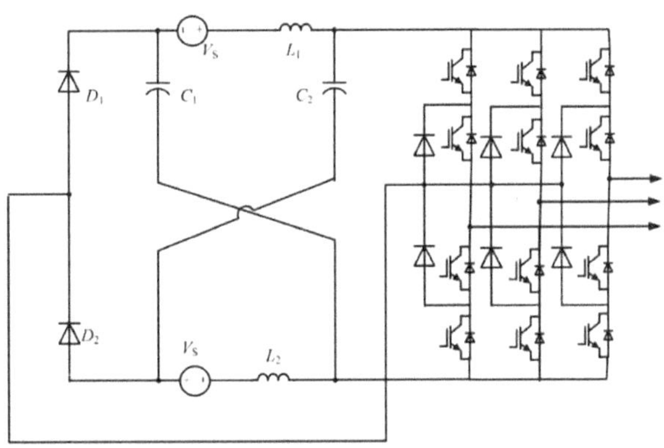

(b) Three-level type

Fig. 3.6 Typical embedded-Z-source converters [18]

Z-source inverters, and its capacitive elements for voltage filtering in current-type embedded-Z-source inverters. The typical topologies can be classified into a two-level type and a three-level type, as shown in Fig. 3.6.

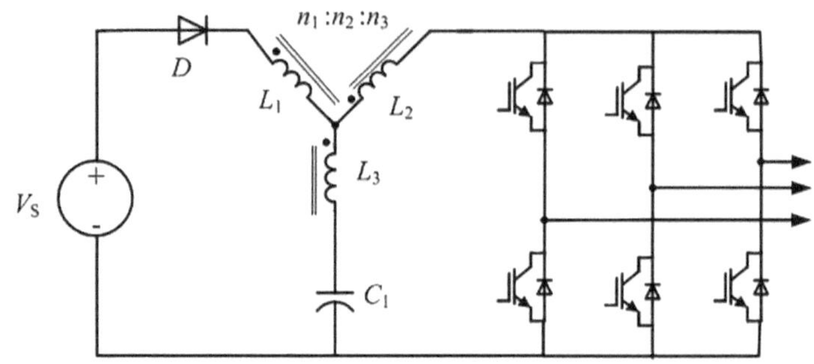

Fig. 3.7 Y-source converters [19]

Fig. 3.8 Γ-Z-source converters [20]

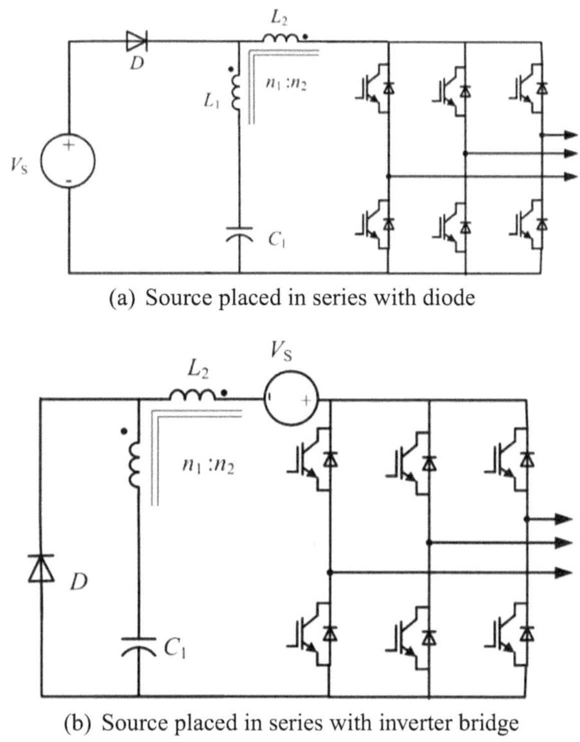

It is remarked that the embedded-Z-source inverters not only maintain the features of typical Z-source inverters, but also generate smaller ripples of input voltage and current.

3.4 Other Impedance Source Converters

In addition to the above-mentioned impedance source converters, some others have also been proposed, like Y-source converters (Fig. 3.7) [19], Γ-Z-source converters (Fig. 3.8) [20–23], LCCT-Z-source converters (Fig. 3.9) [24, 25], and Z-H-source converters (Fig. 3.10) [26], to list just a few.

Y-source converters shown in Fig. 3.7 are designed based on trans-Z-source converters, which, however, realize a higher voltage gain by using a smaller duty ratio.

Γ-Z-source converters shown in Fig. 3.8 use fewer components and a coupled transformer to provide a high voltage gain, and they are essentially derived from the trans-Z-source converters. Therein, two Γ-shaped inductors (Fig. 3.8a) are coupled in trans-Z-source converters to form Γ-Z-source converters. Moreover, a voltage source is embedded in the Γ-shaped network in Fig. 3.8b; therefore, it is also an embedded-Z-source converter.

LCCT-Z-source converters (LCCT stands for the inductor-capacitor-capacitor-transformer) shown in Fig. 3.9 are extended from trans-Z-source inverters and have unique features, such as the converter in Fig. 3.9b, whose two built-in DC blocking capacitors, cascaded with transformer windings, can prevent the transformer from saturation, while the one in Fig. 3.9a, whose built-in DC capacitor, cascaded with

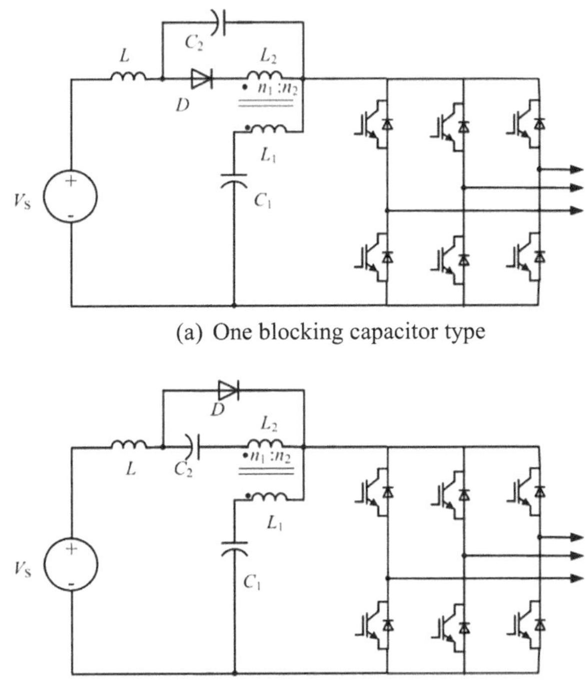

(a) One blocking capacitor type

(b) two blocking capacitors type

Fig. 3.9 LCCT-Z-source converters [24]

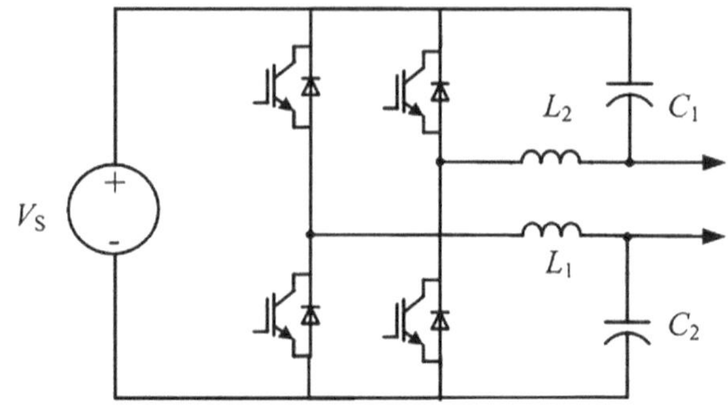

Fig. 3.10 Z-H-source converters [26]

transformer windings, possesses the features of both quasi-Z-source and trans-Z-source converters.

Figure 3.10 depicts a Z-H-source converter, which contains fewer components, but own the same functions as traditional Z-source converters.

Actually, the impedance source techniques have been extended, in detail, the impedance network is placed in other positions of the converter to form impedance networks converters. However, it is remarked that, so far, the design of an impedance networks converter is still an art, and can not fulfil modern industrial requirements. Rapidly developing renewable energy industry has posed more stringent and higher requirements on power electronics, especially high quality converters. Therefore, a systematic designing methodology is desired.

Due to the important role of impedance networks, which are coupled with traditional converters to construct impedance source converters, it is of significance to understand impedance networks matching, which lays a foundation for a systematic designing methodology.

References

1. Y. Siwakoti, F. Peng, F. Blaabjerg, Impedance-Source networks for electric power conversion part I: a topological review. IEEE Trans. Power Electron. **30**(2), 699–716 (Mar. 2015)
2. Y. Siwakoti, F. Peng, F. Blaabjerg, Impedance-Source networks for electric power conversion part II: review of control and modulation techniques. IEEE Trans. Power Electron. **30**(4), 1887–1906 (Apr. 2015)
3. J. Anderson, F.Z. Peng, Four quasi-Z-source inverters, in *IEEE Power Electronics Specialists Conference, 2008. PESC 2008*, vol. 58 no. 1 (2008), pp. 2743–2749
4. M. Azizi, A. Fatemi, M. Mohamadian, A. Yazdian, A novel Z-source four-leg inverter with two independent four-wire outputs, in *Power Electronic and Drive Systems and Technologies Conference (PEDSTC), 2010 1st*, vol. 25 no. 2 (2010), pp. 163–168
5. D. Cao and F.Z. Peng, A family of Z-source and quasi-Z-source DC-DC converters, in *Applied Power Electronics Conference and Exposition, 2009. APEC 2009. Twenty-Fourth Annual IEEE*, vol. 58 no. 1 (2009), pp. 1097–1101
6. F.Z. Peng, M. Shen, Z. Qian, Maximum Boost Control of the Z-source Inverter. IEEE Trans. Power Electron. **20**(4), 833–838 (Jul. 2005)

7. P.C. Loh, D.M. Vilathgamuwa, Y.S. Lai, G.T. Chua, Y. Li, Pulse-Width modulation of Z-source inverters, in *Industry Applications Conference, 2004. 39th IAS Annual Meeting. Conference Record of the 2004 IEEE*, Vol. 1 (2004)
8. P.C. Loh, D.M. Vilathgamuwa, C.J. Gajanayake, Y.R. Lim, C.W. Teo, Transient modeling and analysis of pulse-width modulated Z-source inverter. IEEE Trans. Power Electron. **22**(2), 498–507 (Mar. 2007)
9. C.J. Gajanayake, D.M. Vilathgamuwa, P.C. Loh, Development of a comprehensive model and a multiloop controller for Z-source inverter DG systems. IEEE Trans. Ind. Electron. **54**(4), 2352–2359 (Aug. 2007)
10. P.C. Loh, S.W. Lim, F. Gao, F. Blaabjerg, Three-level Z-source inverters using a single Lc impedance network. IEEE Trans. Power Electron. **22**(2), 706–711 (Mar. 2007)
11. P.C. Loh, D.M. Vilathgamuwa, C.J. Gajanayake, L.T. Wong, C.P. Ang, Z-Source current-type inverters: digital modulation and logic implementation. *Industry Applications Conference, 2005. Fourtieth IAS Annual Meeting. Conference Record of the 2005* (Oct. 2005), pp. 940–947
12. D. Vinnikov, I. Roasto, Quasi-Z-source-based isolated DC-DC converters for distributed power generation. IEEE Trans. Ind. Electron. **58**(1), 192–201 (Jan. 2011)
13. D. Vinnikov, I. Roasto, J. Zakis, R. Strzelecki, New step-up DC-DC converter for fuel cell powered distributed generation systems: some design guidelines. Przeglad Elektrotechniczny **86**, 245–252 (2010)
14. D. Vinnikov, I. Roasto, R. Strzelecki, M.Adamowicz, Step-up DC-DC converters with cascaded quasi-Z-source network. IEEE Trans. Ind. Electron. **59**(10), 3727–3736 (Oct. 2012)
15. D. Vinnikov, I. Roasto, R. Strzelecki, M.Adamowicz, Two-stage quasi-Z-source network based step-up DC-DC converter, in *Industrial Electronics (ISIE), 2010 IEEE International Symposium on.* vol. 58 no. 1 (2010), pp. 1143–1148
16. D. Vinnikov, I. Roasto, Quasi-Z-source-based isolated DC-DC converters for distributed power generation. IEEE Trans. Ind. Electron. **58**(1), 192–201 (Jan. 2011)
17. W. Qian, F.Z. Peng, H. Cha, Trans-Z-source inverters. IEEE Trans. Power Electron. **26**(12), 3453–3463 (Dec. 2011)
18. P.C. Loh, F. Gao, F. Blaabjerg, Embedded EZ-source inverters. IEEE Trans. Ind. Appl. **46**(1), 256–267 (Jan. 2010)
19. Y. Siwakoti, P.C. Loh, F. Blaabjerg, G. Town, Y-source impedance network. IEEE Trans. Power Electron. **29**(7), 3250–3254 (Jul. 2014)
20. P.C. Loh, D. Li, F. Blaabjerg, Γ-Z-Source Inverters. IEEE Trans. Power Electron. Lett. **28**(11), 4880-4884 (Nov. 2013)
21. W. Mo, P.C. Loh, F. Blaabjerg, Asymmetrical Γ-Source Inverters. IEEE Trans. Ind. Electron. **61**(2), 637–647 (Feb. 2014)
22. W. Mo, P.C. Loh, F. Blaabjerg, Voltage type Γ-source inverters with continuous input current and enhanced voltage boost capability, in *15th International Power Electronics and Motion Control Conference and Exposition Europe Congress*, Novi Sad, Serbia, (2012)
23. P.C. Loh, D. Li, F. Blaabjerg, Current-type flipped-Γ-source inverters, in *7th International Power Electronics and Motion Control Conference (IPEMC)*, vol. 1 (2012), pp. 594–598
24. M. Adamowicz, LCCT-Z-Source Inverters, in *10th International Conference on Environment and Electrical Engineering (EEEIC)*, (2011), pp. 1–6
25. M. Adamowicz, R. Strzelecki, F.Z. Peng, J. Guzinski, H. A. Rub, New Type LCCT-Z-Source Inverters, in *Proceedings of the 2011-14th European Conference on Power Electronics and Applications (EPE 2011)*, (2011), pp. 1–10
26. F. Zhang, F.Z. Peng, Z. Qian, Z-H Converter, in *Power Electronics Specialists Conference, 2008. PESC 2008. IEEE*, (2008), pp. 1004–1007

Part II
Impedance Networks Matching

Applying the concept of impedance matching of linear circuits to nonlinear switching circuits leads to impedance network matching, which instructs how an impedance network can be matched to the sources, based on which a systematic design methodology for designing impedance networks converters will be proposed.

Chapter 4
Impedance Networks and Their Matching

Impedance networks matching involves input impedance matching, output impedance matching, and load phase matching, which should be considered together as designing an impedance networks converter.

4.1 Introduction

Power electronic is concerned with the power conversion between various forms and/or characteristics to suit particular applications, so the design of power converters is the most important and basic work.

Power electronics technology is one of the main technologies to realize high-efficiency energy conversion. As is known, there are nearly 70% electric energy should be converted first by power electronics technology to supply energy to the load. Nowadays, power electronics technology has become a basic and key technology in the project of the energy conservation and emission reduction and the project of the green energy [1].

There are three main research directions in power electronics technology, i.e. research of power switches components, novel power converters topologies, and control strategies. Therein, the research of novel power converters topologies is consistently the hot research topic at home and abroad [2, 3].

In recent three decades, thousands of different kinds of converters are proposed to satisfy the requirements of the high-efficiency energy conversion [4, 5], e.g. soft-switching converters, pulse-width-modulation (short for PWM) rectifiers, multi-level converters, high-gain DC-DC converters and Z-source converters, etc.

The soft-switching converters were proposed to solve the problems caused by the high-frequency, e.g. the high switching energy loss, high EMI problems [6, 7]. Therein, additional inductors and capacitors were utilized to realize the zero-current switching-off or zero-voltage switching-on of the power switches by resonance, and

© Springer International Publishing AG 2018

G. Zhang et al., *Designing Impedance Networks Converters*, Studies in Systems, Decision and Control 119, DOI 10.1007/978-3-319-63655-9_4

then to reduce the switching loss, increase the efficiency and reduce EMI of the converter [8, 9].

PWM rectifiers were invented to solve the harmonic and reactive pollution caused by uncontrolled rectifiers. The main improvement is that the controlled switches are utilized to instead of the uncontrolled ones in uncontrolled rectifiers. Thereafter, PWM technology was applied in it to make the current of the power grid to be a sine wave, and then the converter operates in an ideal power factor [10, 11]. PWM rectifiers can be divided into two kinds, i.e. voltage-type and current-type rectifiers. Thereby, a lot of topologies were derived from these two kinds of rectifiers, which make a great reduction in the power grid pollution and increase the efficiency [12, 13].

In order to realize the power conversion in medium, high AC driving voltage and the flexible AC transmission systems, multilevel converters were developed [14, 15]. Multilevel converters were derived from the one proposed by A. Nabae [16], i.e. the diode-clamp type multilevel converters and the flying-capacitor multilevel converters. Therein, an H-bridge cascaded multilevel converter is the typical one, which was composed by several cascaded H-bridge multilevel converter, moreover, which can not only realize the high power-level output, but also make it as a module and easy to be expanded and repaired [15]. In fact, modulation multilevel converters (short for MMC) were also derived from the cascaded multilevel converter [17].

In order to boost the low voltage of solar energy systems and fuel battery systems to match the high bus voltage, high-gain DC-DC converters with high output voltage were required [18–21]. High-gain DC-DC converters were derived from boost DC-DC converters, and then boost technologies, i.e. cascaded boost transformers, coupled inductors, and switching capacitors and inductors, were applied in boost converters to form different kinds of high-gain DC-DC boost converters [22–25].

Z-source converters were proposed to overcome the disadvantages of traditional voltage-source inverters, i.e. the shoot-through problems in the same bridge, the inability of loading a capacitor load, and low voltage of the DC output port [26]. Therein, an LC impedance network was located in the DC input port of the inverter, and then the improve inverter obtains some unique features, which promote the Z-source inverter owes the application and development prospects [27, 28].

Throughout the developments and investigation process of power converters topologies, it is found that nearly all the converters are mainly designed and invented by the researcher's experiences or inspirations, it lacks of a systematic and theoretical guidance to design a power converter topology. With the development of power electronics, the traditional research method obviously cannot satisfy the requirements of practical applications. Therefore, it is of great significance to develop a design methodology of power converters topologies. The topologies of impedance networks converters are widely studied and applied due to its unique features and its constructing method, i.e. an impedance network is applied to couple with the traditional converters to improve their features. The design method of Z-source converters is essentially a methodology developed from the impedance networks matching, which, however, is not in-depth analysis of the impedance matching features, moreover, which has not yet been revealed and formed a systematic methodology to design impedance networks converters. Hence, the proposal of a systematic design methodology for impedance networks converters would play a great rôle in the development of power converters design and constructions.

4.2 Impedance Matching

The concept of impedance matching deals originally with linear circuits, and is not directly applicable to power converters, which are essentially nonlinear switching circuits. Nevertheless, in each operational mode, a power converter works as a linear circuit, which results in the time-varying characteristics of the impedance networks matching. Therefore, the concept of the impedance matching can be extended to the impedance networks matching in three aspects: input impedance matching, output impedance matching, and load phase matching.

4.3 Input Impedance Matching

Substituting $s = j\omega$ into the input impedance of the two-port network in (2.5) results in

$$Z_{\mathrm{i}}(j\omega) = \mathrm{Re}\left(\frac{A_{11}(j\omega)Z_{\mathrm{L}}(j\omega) + A_{12}(j\omega)}{A_{21}(j\omega)Z_{\mathrm{L}}(j\omega) + A_{22}(j\omega)}\right) + j\mathrm{Im}\left(\frac{A_{11}(j\omega)Z_{\mathrm{L}}(j\omega) + A_{12}(j\omega)}{A_{21}(j\omega)Z_{\mathrm{L}}(j\omega) + A_{22}(j\omega)}\right). \quad (4.1)$$

The shoot-through state implies that $Z_{\mathrm{L}}(j\omega) = 0$, so the input impedance in shoot-through state is derived as

$$Z_{\mathrm{i}}(j\omega) = \mathrm{Re}\left(\frac{A_{12}(j\omega)}{A_{22}(j\omega)}\right) + j\mathrm{Im}\left(\frac{A_{12}(j\omega)}{A_{22}(j\omega)}\right), \quad (4.2)$$

whereas the input current of the voltage source at the shoot-through states is expressed as

$$I_{\mathrm{S}}(j\omega) = \frac{V_{\mathrm{S}}(j\omega)}{Z_{\mathrm{i}}(j\omega)} = \frac{V_{\mathrm{S}}(j\omega)}{\mathrm{Re}\left(\frac{A_{12}(j\omega)}{A_{22}(j\omega)}\right) + j\mathrm{Im}\left(\frac{A_{12}(j\omega)}{A_{22}(j\omega)}\right)}. \quad (4.3)$$

Since inductive components hinder their current change, it is then obvious that the converter can restrain the short-circuited current if its input impedance in (4.2) is inductive.

$$\mathrm{Im}\left(\frac{A_{12}(j\omega)}{A_{22}(j\omega)}\right) > 0. \quad (4.4)$$

4.4 Output Impedance Matching

Substituting $s = j\omega$ into the output voltage equation $V_L(s) = V_S(s) - I_L(s)Z_o(s)$ results in $V_L(j\omega) = V_S(j\omega) - I_L(j\omega)Z_o(j\omega)$. It is remarked that in order for the output voltage to be higher than the source voltage, the output impedance $Z_o(j\omega)$ should be negative; otherwise, the output impedance $Z_o(j\omega)$ should be positive.

Substituting $s = j\omega$ into (2.6) leads to the output impedance of the two-port network as

$$Z_o(j\omega) = \mathrm{Re}\left(\frac{A_{22}(j\omega)Z_S(j\omega) + A_{12}(j\omega)}{A_{21}(j\omega)Z_S(j\omega) + A_{11}(j\omega)}\right) + j\mathrm{Im}\left(\frac{A_{22}(j\omega)Z_S(j\omega) + A_{12}(j\omega)}{A_{21}(j\omega)Z_S(j\omega) + A_{11}(j\omega)}\right), \quad (4.5)$$

while the corresponding output voltage is

$$V_L(j\omega) = V_S(j\omega)\frac{Z_L(j\omega)}{Z_L(j\omega) + Z_o(j\omega)}. \quad (4.6)$$

It is obvious that $|V_L(j\omega)| > |V_S(j\omega)|$, if the voltage gain M satisfies the condition $M > 1$, namely,

$$\begin{aligned} M &= \frac{|V_L(j\omega)|}{|V_S(j\omega)|} \\ &= |\frac{Z_L(j\omega)}{Z_L(j\omega) + Z_o(j\omega)}| \\ &= \frac{|Z_L(j\omega)|}{|Z_L(j\omega) + Z_o(j\omega)|} \\ &= \frac{\sqrt{[\mathrm{Re}(Z_L(j\omega))]^2 + [\mathrm{Im}(Z_L(j\omega))]^2}}{\sqrt{[\mathrm{Re}(Z_L(j\omega)) + \mathrm{Re}(Z_o(j\omega))]^2 + [\mathrm{Im}(Z_L(j\omega)) + \mathrm{Im}(Z_o(j\omega))]^2}} \\ &> 1, \end{aligned} \quad (4.7)$$

from which one has

$$[\mathrm{Re}(Z_L(j\omega))]^2 + [\mathrm{Im}(Z_L(j\omega))]^2 > [\mathrm{Re}(Z_L(j\omega)) + \mathrm{Re}(Z_o(j\omega))]^2 + [\mathrm{Im}(Z_L(j\omega)) + \mathrm{Im}(Z_o(j\omega))]^2, \quad (4.8)$$

which can be further simplified as

$$\begin{aligned} &2[\mathrm{Re}(Z_L(j\omega))\mathrm{Re}(Z_o(j\omega)) + \mathrm{Im}(Z_L(j\omega))\mathrm{Im}(Z_o(j\omega))] + [\mathrm{Re}(Z_o(j\omega))]^2 + [\mathrm{Im}(Z_o(j\omega))]^2 \\ &= 2[\mathrm{Re}(Z_L(j\omega))\mathrm{Re}(Z_o(j\omega)) + \mathrm{Im}(Z_L(j\omega))\mathrm{Im}(Z_o(j\omega))] + |Z_o(j\omega)|^2 \\ &< 0. \end{aligned} \quad (4.9)$$

Then, if one has

$$\mathrm{Re}(Z_L(j\omega))\mathrm{Re}(Z_o(j\omega)) + \mathrm{Im}(Z_L(j\omega))\mathrm{Im}(Z_o(j\omega)) < -\frac{|Z_o(j\omega)|^2}{2} < 0, \quad (4.10)$$

(4.7) holds. That is, if $\mathrm{Re}(Z_L(j\omega))\mathrm{Re}(Z_o(j\omega) < 0$ or $\mathrm{Im}(Z_L(j\omega))\mathrm{Im}(Z_o(j\omega)) < 0$, and their sum is smaller than 0, then (4.10) holds. Moreover, it is suggested from (4.10) that the real parts of the load impedance and the output impedance should have opposite signs, or the imaginary parts of the load impedance and the output impedance should have opposite signs. This means that the output impedance should have negative impedance features; otherwise, the output impedance exhibits positive impedance features.

4.5 Load Phase Matching

In order to improve the load ability of the converter, so that the inverter is applicable to both inductive and capacitive loads, the output impedance phase of the inverter should be capacitive or inductive so as to match the load impedance for reducing the impedance phase angle of the inverter. Therein, the total impedance phase is the sum of the output impedance phase and load impedance phase. Moreover, the smaller the total impedance phase is, the larger the power factor of the inverter is. Therefore, the optimal condition is that its total impedance phase is 0°.

The impedance phase angle of the converter is given by

$$\varphi = \arctan\left(\frac{\mathrm{Im}\,(Z_o(j\omega) + Z_L(j\omega))}{\mathrm{Re}\,(Z_o(j\omega) + Z_L(j\omega))}\right). \tag{4.11}$$

In terms of (4.5),

$$\mathrm{Im}\,(Z_o(j\omega) + Z_L(j\omega)) = \mathrm{Im}\left(\frac{A_{22}(j\omega)Z_S(j\omega) + A_{12}(j\omega)}{A_{21}(j\omega)Z_S(j\omega) + A_{11}(j\omega)} + Z_L(j\omega)\right) = 0 \tag{4.12}$$

implies that its impedance phase angle is 0°. Moreover, (4.12) also can be further simplified to

$$\mathrm{Im}\,(Z_L(j\omega)) = -\mathrm{Im}\,(Z_o(j\omega)) = -\mathrm{Im}\left(\frac{A_{22}(j\omega)Z_S(j\omega) + A_{12}(j\omega)}{A_{21}(j\omega)Z_S(j\omega) + A_{11}(j\omega)}\right). \tag{4.13}$$

4.6 Matching Practice

It is now known that input impedance matching is to increase the input impedance in the short-circuited case for making the input impedance inductive and then to restrain the input current; output impedance matching is to tune the output impedance to be of positive or negative nature, so as to increase or decrease output voltage by connecting an impedance network or adjusting the impedance networks parameters; while load phase matching is to match the output impedance with the load impedance to ensure its impedance phase angle to be 0°.

Therefore, to design a reasonable and feasible impedance network converter, input impedance matching, output impedance matching and load phase matching should be considered together.

From Sects. 4.3, 4.4 and 4.5, conditions for impedance networks matching can be concluded as

$$\begin{cases} \mathrm{Im}\left(\dfrac{A_{12}(j\omega)}{A_{22}(j\omega)}\right) > 0\,, \\ \mathrm{Re}(Z_{\mathrm{L}}(j\omega))\mathrm{Re}(Z_{\mathrm{o}}(j\omega)) + \mathrm{Im}(Z_{\mathrm{L}}(j\omega))\mathrm{Im}(Z_{\mathrm{o}}(j\omega)) < -\dfrac{|Z_{\mathrm{o}}(j\omega)|^2}{2}\,, \\ \mathrm{Im}\left(Z_{\mathrm{L}}(j\omega)\right) = -\mathrm{Im}\left(\dfrac{A_{22}(j\omega)Z_{\mathrm{S}}(j\omega) + A_{12}(j\omega)}{A_{21}(j\omega)Z_{\mathrm{S}}(j\omega) + A_{11}(j\omega)}\right). \end{cases} \tag{4.14}$$

To satisfy the conditions (4.14), it is concerned with the topologies and parameters of impedance networks, the source impedance, and the load impedance; while the matching process is to calculate the parameters of impedance networks in terms of (4.14) and other known parameters, which are to be concretized in the next section.

References

1. J.D. van Wyk, F.C. Lee, On a future for power electronics. IEEE Trans. Emerg. Sel. Top. Power Electron. **1**(2), 59–72 (2013)
2. D. Cao, F.Z. Peng, A family of Z-source and quasi-Z-source DC-DC converters, in *Applied Power Electronics Conference and Exposition, 2009. APEC 2009. Twenty-Fourth Annual IEEE*, vol. 58, no. 1 (2009) pp. 1097–1101
3. B.K. Bose, *Modern Power Electronics and AC Drives*, vol. 123 (Prentice Hall, USA, 2002)
4. J.A. Baroudi, V. Dinavahi, A.M. Knight, A review of power converter topologies for wind generators. Renew. Energy **32**(14), 2369–2385 (2007)
5. P.W. Wheeler, J. Rodriguez, J.C. Clare, Matrix converters: a technology review. IEEE Trans. Ind. Electron. **49**(2), 276–288 (2002)
6. B. Zhang, D. Qiu, *Sneak Circuits of Power Electronic Converters* (John Wiley and Sons, 2014)
7. G. Hua, F.C. Lee, Soft-switching techniques in PWM converters. IEEE Trans. Ind. Electron. **42**(6), 595–603 (1995)
8. C. Mario, C. Alfio, A. Rosario, G. Francesco, Soft-switching converter with HF transformer for grid-connected photovoltaic systems. IEEE Trans. Ind. Electron. **57**(5), 1678–1686 (2010)
9. M.D. Bellar, T.S. Wu, A. Tchamdjou, A review of soft-switched DC-AC converters. IEEE Trans. Ind. Electron. **34**(4), 847–860 (1998)
10. H. Mao, F.C. Lee, D. Boroyevich, S. Hiti, Review of high-performance three-phase power-factor correction circuits. IEEE Trans. Ind. Electron. **44**(4), 437–446 (1997)
11. B. Singh, B.N. Singh, A. Chandra, K. Al-Haddad, A. Pandey, D.P. Kothari, A review of single-phase improved power quality AC-DC converters. IEEE Trans. Ind. Electron. **50**(5), 962–981 (2003)

12. J.R. Rodríguez, L.W. Dixon, J.R. Espinoza, J. Pontt, P. Lezana, PWM regenerative rectifiers: state of the art. IEEE Trans. Ind. Electron. **52**(1), 5–22 (2005)
13. J.W. Kolar, T. Friedli, J.R. Rodríguez, P.W. Wheeler, Review of three-phase PWM AC-AC converter topologies. IEEE Trans. Ind. Electron. **58**(11), 4988–5006 (2011)
14. L.G. Franquelo, J. Rodriguez, J.I. Leon, S. Kouro, R. Portillo, M.A. Prats, The age of multilevel converters arrives. IEEE Ind. Electron. Mag. **2**(2), 28–39 (2008)
15. I. Colak, E. Kabalci, R. Bayindir, Review of multilevel voltage source inverter topologies and control schemes. Energy Convers. Manag. **52**(2), 1114–1128 (2011)
16. A. Nabae, I. Takahashi, H. Akagi, A new neutral-point-clamped PWM inverter. IEEE Trans. Ind. Appl. **5**, 518–523 (1981)
17. S. Rohner, S. Bernet, M. Hiller, R. Sommer, Modulation, Losses, and Semiconductor Requirements of Modular Multilevel Converters. IEEE Trans. Ind. Electron. **57**(8), 2633–2642 (2010)
18. W. Li, J. Liu, J. Wu, X. He, Design and analysis of isolated ZVT boost converters for high-efficiency and high-step-up applications. IEEE Trans. Power Electron. **22**(6), 2263–2374 (2007)
19. W. Li, X. He, A family of interleaved DC-DC converter deduced from a basic cell with winding-cross coupled inductors. IEEE Trans. Power Electron. **23**(4), 1791–1801 (2008)
20. H.B. Shin, J.G. Park, S.K. Chung, H.W. Lee, T.A. Lipo, Generalised steady-state analysis of multiphase interleaved boost converter with couple inductors. IEE Proc. Electr. Power Appl. **152**(3), 584–594 (2005)
21. W.C. Li, X. Xiang, C.S. Li, W.H. Li, X. He, Interleaved high step-up ZVT converter with built-in transformer voltage doubler cell for distributed PV generation system. IEEE Trans. Power Electron. **28**(1), 300–313 (2013)
22. Y. Gu, W. Li, Y. Zhao, B. Yang, C. Li, X. He, Transformerless inverter with virtual DC bus concept for cost-effective grid-connected PV power systems. IEEE Trans. Power Electron. **28**(2), 793–805 (2013)
23. Y. Zhao, X. Xiang, W. Li, X. He, C. Xia, Advanced symmetrical voltage quadrupler rectifiers for high step-up and high output-voltage converters. IEEE Trans. Power Electron. **28**(4), 1622–1631 (2013)
24. W. Li, Y. Zhao, J. Wu, X. He, Interleaved high step-up converter with winding-cross-coupled inductors and voltage multiplier cells. IEEE Trans. Power Electron. **27**(1), 133–143 (2012)
25. W. Li, L. Fan, Y. Zhao, X. He, D. Xu, High step-up and high efficiency fuel cell power generation system with active clamp flyback-forward converter. IEEE Trans. Ind. Electron. **59**(1), 599–610 (2012)
26. F.Z. Peng, Z-source inverter. IEEE Trans. Ind. Appl. **39**(2), 504–510 (2003)
27. Y. Siwakoti, F. Peng, F. Blaabjerg, Impedance-source networks for electric power conversion part I: a topological review. IEEE Trans. Power Electron. **30**(2), 699–716 (2015)
28. Y. Siwakoti, F. Peng, F. Blaabjerg, Impedance-source networks for electric power conversion part II: review of control and modulation techniques. IEEE Trans. Power Electron. **30**(4), 1887–1906 (2015)

Chapter 5
Design Methodology of Impedance Networks Converters

In terms of the impedance networks matching, a systematic design methodology of impedance networks converters is presented in this chapter.

5.1 Topology Design

The topology of impedance networks determines the performance of converters, which deals with impedance networks connectivity and components location.

5.1.1 Impedance Networks Connectivity

The impedance networks connectivity refers to the terminals connectivity of two impedance networks. There are four terminals in each impedance network, which results in different connectivity types of two-port networks, i.e. cascade, parallel, series-parallel, parallel-series, and series-series connectivity, as well corresponding operational rules, as shown in Table 5.1.

Therein, **A**, **Y**, **H**, **G**, and **Z** are the transmission parameters matrix, admittance parameters matrix, hybrid parameters matrix, inverse hybrid parameters matrix and impedance parameters matrix, respectively. Moreover, the connectivity of two-port networks shown in Table 5.1 can be the connectivity between two impedance networks, between an impedance network and the source, or between an impedance network and a load.

© Springer International Publishing AG 2018

G. Zhang et al., *Designing Impedance Networks Converters*, Studies in Systems, Decision and Control 119, DOI 10.1007/978-3-319-63655-9_5

Table 5.1 Connectivity types of two-port networks and the operational rules

Type	Connectivity	Operations
Cascaded	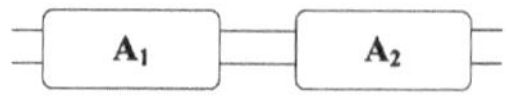	$\mathbf{A} = \mathbf{A}_1\mathbf{A}_2$
Parallelled	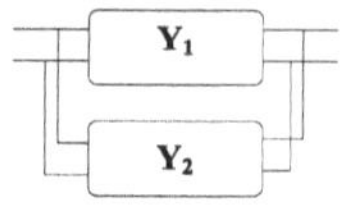	$\mathbf{Y} = \mathbf{Y}_1 + \mathbf{Y}_2$
Series-parallel	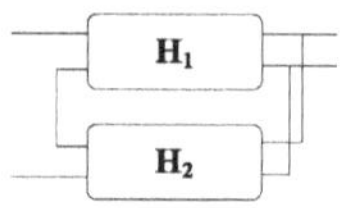	$\mathbf{H} = \mathbf{H}_1 + \mathbf{H}_2$
Parallel-series	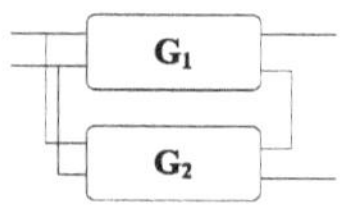	$\mathbf{G} = \mathbf{G}_1 + \mathbf{G}_2$
Series-series	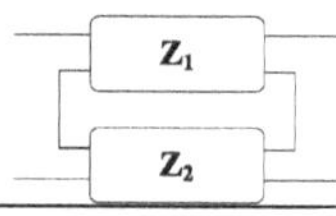	$\mathbf{Z} = \mathbf{Z}_1 + \mathbf{Z}_2$

5.1.2 Location of the Impedance Networks

The location of an impedance network can be varied.

1. As shown in Fig. 5.1a, the impedance network can be located between the source and a part of the original converter, e.g. the converters in Figs. 2.7, 2.5a, 3.1, 3.4, 3.7, 3.8a, and 3.9.
2. As shown in Fig. 5.1b, the impedance network can be located inside the original converter, e.g. the Z-source AC-AC converter in Fig. 2.5c.
3. As shown in Fig. 5.1c, the impedance network can be located between the original converter and the load, e.g. the Z-source AC-DC converter in Fig. 2.5b and the Z-H-source converter in Fig. 3.10.

5.2 Selection of Impedance Networks

Basic two-port impedance networks and their transmission parameters matrices are listed in Table 5.2, where $Z_x = Z_1Z_2Z_3Z_4$.

Denote the source impedance in the input port of the two-port network as Z_{2S} and the load impedance in the output port of the two-port network as Z_{2L}, and the corresponding input and output impedances can be expressed as given in Table 5.3 in

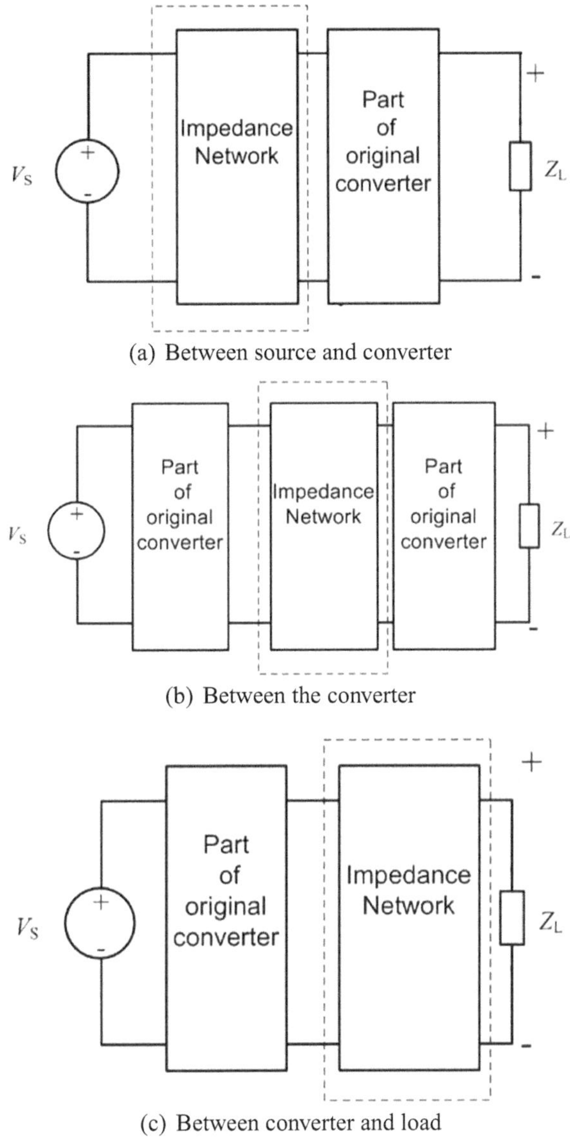

Fig. 5.1 Location of an impedance network in converters

terms of (2.5) and (2.6). Moreover, the impedance phase angle of two-port impedance networks is given as

$$\varphi = \arctan\left(\frac{\mathrm{Im}\,(Z_o + Z_{2L})}{\mathrm{Re}\,(Z_o + Z_{2L})}\right). \tag{5.1}$$

Table 5.2 Basic two-port impedance networks and their transmission parameters matrices

Networks type	Figure	Transmission matrix **A**
Basic cascaded	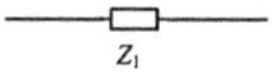	$\begin{bmatrix} 1 & Z_1 \\ 0 & 1 \end{bmatrix}$
Basic parallelled	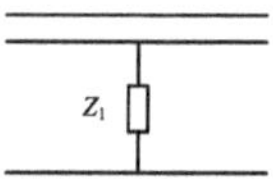	$\begin{bmatrix} 1 & 0 \\ \frac{1}{Z_1} & 1 \end{bmatrix}$
X-shaped	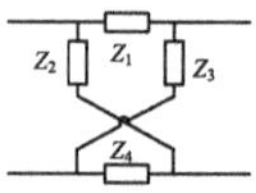	$\begin{bmatrix} \frac{Z_x+Z_1Z_3+Z_2Z_4+1}{1-Z_x} & \frac{Z_1Z_2Z_4+Z_1Z_3Z_4+Z_1+Z_4}{1-Z_x} \\ \frac{Z_1Z_2Z_3+Z_2Z_3Z_4+Z_2+Z_3}{1-Z_x} & \frac{Z_x+Z_1Z_2+Z_3Z_4+1}{1-Z_x} \end{bmatrix}$
π-shaped	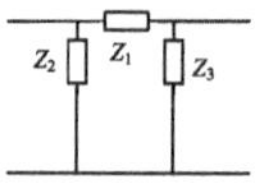	$\begin{bmatrix} \frac{Z_1+Z_3}{Z_3} & Z_1 \\ \frac{Z_1+Z_2+Z_3}{Z_2Z_3} & \frac{Z_1+Z_2}{Z_2} \end{bmatrix}$
Γ-shaped	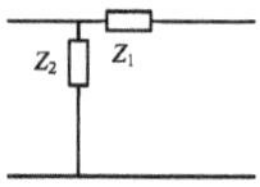	$\begin{bmatrix} 1 & Z_1 \\ \frac{1}{Z_2} & \frac{Z_1+Z_2}{Z_2} \end{bmatrix}$
Inverse-Γ-shaped	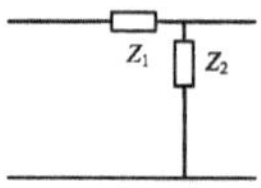	$\begin{bmatrix} \frac{Z_1+Z_2}{Z_2} & Z_1 \\ \frac{1}{Z_2} & 1 \end{bmatrix}$
T-shaped	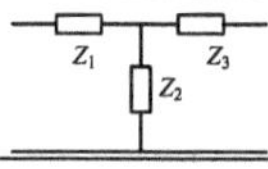	$\begin{bmatrix} \frac{Z_1+Z_2}{Z_2} & \frac{Z_1Z_2+Z_1Z_3+Z_2Z_3}{Z_2} \\ \frac{1}{Z_2} & \frac{Z_3+Z_2}{Z_2} \end{bmatrix}$

Table 5.3 Input and output impedances of two-port networks

Networks types	Z_i	Z_o
Basic cascaded	$Z_{2L}+Z_1$	$Z_{2S}+Z_1$
Basic parallelled	$\frac{Z_{2L}Z_1}{Z_{2L}+Z_1}$	$\frac{Z_{2S}Z_1}{Z_{2S}+Z_1}$
X-shaped	Z_{Xi}	Z_{Xo}
π-shaped	$\frac{(Z_1+Z_3)Z_2Z_{2L}+Z_1Z_2Z_3}{(Z_1+Z_2+Z_3)Z_{2L}+(Z_1+Z_2)Z_3}$	$\frac{(Z_1+Z_2)Z_3Z_{2S}+Z_1Z_2Z_3}{(Z_1+Z_2+Z_3)Z_{2S}+(Z_1+Z_3)Z_2}$
Γ-shaped	$\frac{Z_2Z_{2L}+Z_1Z_2}{Z_{2L}+Z_1+Z_2}$	$\frac{(Z_1+Z_2)Z_{2S}+Z_1}{Z_{2S}+Z_+Z_2}$
Inverse-Γ-shaped	$\frac{(Z_1+Z_2)Z_{2L}+Z_1}{Z_{2L}+Z_+Z_2}$	$\frac{Z_2Z_{2S}+Z_1Z_2}{Z_{2S}+Z_1+Z_2}$
T-shaped	$\frac{(Z_1+Z_2)Z_{2L}+Z_T}{Z_{2L}+Z_2+Z_3}$	$\frac{(Z_3+Z_2)Z_{2S}+Z_T}{Z_{2S}+Z_2+Z_1}$

Substituting the output impedance Z_o in Table 5.3 into (5.1) results in the corresponding impedance phase angles, which helps selecting proper impedance networks.

It is remarked that

1. $Z_{Xi} = \dfrac{(Z_x + Z_1Z_3 + Z_2Z_4 + 1)Z_{2L} + Z_1Z_2Z_4 + Z_1Z_3Z_4 + Z_1 + Z_4}{(Z_1Z_2Z_3 + Z_2Z_3Z_4 + Z_2 + Z_3)Z_{2L} + Z_1Z_2 + Z_3Z_4 + Z_x + 1}$,
2. $Z_{Xo} = \dfrac{(Z_x + Z_1Z_2 + Z_3Z_4 + 1)Z_{2S} + Z_1Z_2Z_4 + Z_1Z_3Z_4 + Z_1 + Z_4}{(Z_1Z_2Z_3 + Z_2Z_3Z_4 + Z_2 + Z_3)Z_{2S} + Z_1Z_3 + Z_2Z_4 + Z_x + 1}$,
3. $Z_T = Z_1Z_2 + Z_1Z_3 + Z_2Z_3$.

5.3 Input Impedance

In different operational modes, a power converter works as different linear circuits, which leads to the time-varying characteristics of impedance matching. Three cases of power converters, i.e. short-circuited, open-circuit, and normal operational cases, correspond to three input impedance cases of a two-port network. In short-circuited case, the output port of the two-port network is short-circuited, and its input impedance is $Z_{is} = Z_i|_{Z_{2L}=0}$. Similarly, the open-circuit case refers to the case that the output port of the two-port network is open-circuit, and its input impedance writes $Z_{io} = Z_i|_{Z_{2L}=\infty}$. Further, the input impedance of a two-network in normal case refers to $Z_{ic} = Z_i$. Finally, the input impedance in various cases is summarized in Table 5.4.

5.3.1 *Short-Circuited Case*

According to Table 5.4, it is straightforward that the input impedance of the basic parallel type in short-circuited case is 0, implying that it lacks the ability of preventing from short-circuited case; while the basic cascade type also cannot prevent from the short-circuited case, when Z_1 is very small, i.e. the voltage source inverter is also a typical cascade type but it cannot prevent from the short-circuited case because Z_1 is close to 0.

5.3.2 *Open-Circuit Case*

According to Table 5.4, it is obvious that the open-circuit input impedance of the basic cascaded type is infinite, implying that it lacks of the ability preventing the open-circuit case in the load, while the basic parallelled type also cannot prevent the open-circuit case when $Z_1 = \infty$, i.e. the current source inverter.

Table 5.4 Input impedances of two-port networks in different cases

Networks types	Short-circuited case Z_{is}	Open-circuit case Z_{io}
Basic cascaded	Z_1	∞
Basic parallelled	0	Z_1
X-shaped	$\frac{Z_1Z_2Z_4+Z_1Z_3Z_4+Z_1+Z_4}{Z_1Z_2+Z_3Z_4+Z_1Z_2Z_3Z_4+1}$	$\frac{Z_1Z_3+Z_2Z_4+Z_1Z_2Z_3Z_4+1}{Z_1Z_2Z_3+Z_2Z_3Z_4+Z_2+Z_3}$
π-shaped	$\frac{Z_1Z_2}{Z_1+Z_2}$	$\frac{(Z_1+Z_3)Z_2}{Z_1+Z_2+Z_3}$
Γ-shaped	$\frac{Z_1Z_2}{Z_1+Z_2}$	Z_2
Inverse-Γ-shaped	$\frac{Z_1}{Z_2}$	$Z_1 + Z_2$
T-shaped	$\frac{Z_1Z_2+Z_1Z_3+Z_2Z_3}{Z_2+Z_3}$	$Z_1 + Z_2$

5.3.3 Normal Case

When a converter operates in a normal case, its input impedance of each two-port network is given in Table 5.3.

Normally, one can obtain the average input impedance in one switching period with the corresponding control strategy as

$$Z_{\mathrm{i}} = d_1 Z_{\mathrm{is}} + d_2 Z_{\mathrm{io}} + (1 - d_1 - d_2) Z_{\mathrm{ic}} \,, \tag{5.2}$$

where d_1, d_2, and $1 - d_1 - d_2$ are the corresponding duty cycles of input impedances in three cases, respectively.

5.3.4 Output Impedance

Similarly, the output impedances of the two-port network can be obtained in three cases as follows.

Case 1: $Z_{2\mathrm{S}} = 0$

Suppose that the input port of the two-port network is connected with an ideal voltage source, i.e. $Z_{\mathrm{os}} = Z_{\mathrm{o}}|_{Z_{2\mathrm{S}}=0}$. Substituting $Z_{2\mathrm{S}} = 0$ into the output impedance equations in Table 5.3 results in the corresponding output impedances in Table 5.5.

In terms of Table 5.5 that the output impedance of the basic cascade type in this case is Z_1, and Z_1 determines the features of the output impedance, which leads to low output voltage, e.g. the voltage source inverter in Fig. 2.3a, where $Z_1 = Z_{\mathrm{S}}$. It is obvious that the output voltage is lower than the input voltage due to the output impedance, and it can only load the inductive loads.

Case 2: $Z_{2\mathrm{S}} \to \infty$

Suppose that the input port of the two-port network is open-circuit, i.e. $Z_{\mathrm{oo}} = Z_{\mathrm{o}}|_{Z_{2\mathrm{S}}=\infty}$. Substituting $Z_{2\mathrm{S}} = \infty$ into the output impedance equations in Table 5.3 leads to the corresponding output impedances, given in Table 5.5.

Table 5.5 Output impedances of two-port networks in different cases

Networks type	$Z_{2\mathrm{S}} = 0$	$Z_{2\mathrm{S}} \to \infty$
Basic cascaded	Z_1	$\to \infty$
Basic parallelled	0	Z_1
X-shaped	$\frac{Z_1 Z_2 Z_4 + Z_1 Z_3 Z_4 + Z_1 + Z_4}{Z_1 Z_3 + Z_2 Z_4 + Z_1 Z_2 Z_3 Z_4 + 1}$	$\frac{Z_1 Z_2 + Z_3 Z_4 + Z_1 Z_2 Z_3 Z_4 + 1}{Z_1 Z_2 Z_3 + Z_2 Z_3 Z_4 + Z_2 + Z_3}$
π-shaped	$\frac{Z_1 + Z_3}{Z_1 + Z_3}$	$\frac{(Z_1 + Z_2) Z_3}{Z_1 + Z_2 + Z_3}$
Γ-shaped	$\frac{Z_1}{Z_2}$	$Z_1 + Z_2$
Inverse-Γ-shaped	$\frac{Z_1 Z_2}{Z_1 + Z_2}$	Z_2
T-shaped	$\frac{Z_1 Z_2 + Z_1 Z_3 + Z_2 Z_3}{Z_2 + Z_1}$	$Z_3 + Z_2$

Case 3: Normal Case

Denote the output impedance in the normal case as Z_{oc} in Table 5.3.

It is remarked that the average output impedance in one switching period is given by

$$Z_o = d_3 Z_{os} + d_4 Z_{oo} + (1 - d_3 - d_4) Z_{oc} \,, \tag{5.3}$$

where d_3, d_4, and $1 - d_3 - d_4$ are the corresponding duty cycles of input impedances in three cases, respectively.

Similarly, the impedance phase of a converter can be obtained according to Table 5.5 and (5.1). Then, the input impedance matching, output impedance matching, and load phase matching are comprehensively analysed to realize optimized matching.

5.4 Analysis of the Operational Modes

Power switches in a power converter lead to different kinds of operational modes, which should be analysed in detail to understand the overall performances of the converter. The analysis includes the energy transfer process and deductions of voltage and current relationships through the law of conservation of energy, Kirchhoff's current and voltage laws (short for KCL and KVL), and other basic circuit laws.

5.5 Parameters Determination

According to section and the impedance matching conditions in (4.14), the parameters of the converter are determined via the transmission parameters matrix.

5.6 Simulations and Experiments

According to the parameters obtained in Sect. 5.5, simulations are conducted to verify the designed converters via circuit simulations softwares, e.g. MatLab-Simulink, PSIM and PSPICE.

Part III
Designing Novel Impedance Networks Converters

In terms of the proposed design methodology, several impedance networks converters are to be designed for specific renewable energy applications.

Chapter 6
3-Z-Network Boost Converter

Two novel boost converters with three active Z-networks are to be designed in this chapter. Therein, impedance networks are cascaded to realize impedance matching, i.e. the output impedance is matched with negative impedance features for realizing high voltage gains. The proposed 3-Z-network boost converters can realize 350 of voltage gain in theory using only one switch, and it can well fulfill the stringent requirements from industry, especially renewable power systems, to boost low voltage from clean sources like photovoltaic (PV) arrays and fuel cells to high voltages for grid-connected converters.

6.1 Introduction

Modern industry, especially renewable energy industry like photovoltaic arrays and fuel cells [1], has posed quite high requirements on power electronics technology [2–5]. For instance, high-step-up DC-DC converters have been used to boost low voltages (18–56 V) to high voltages (200–400 V) for feeding into grid-connected inverters [6, 7]. Further, high-step-up and high-efficiency converters are also required in many other industrial applications, e.g. high-intensity discharge lamps for automobile headlights where the battery voltage of 12 V must raise up to 100 V [8], back-up energy conversions like non-interruptible power systems where the low battery voltage of 48 V must be boosted up to 380 V [9, 10], front-end stages of electric vehicles [11] and communication power systems [12, 13], and some other special applications like electroplating [14].

For those purposes, traditional boost converters or flyback converters have been normally adopted [15–17]. However, conventional boost converters are restricted by parasitic effects of their components and serious energy loss. Moreover, it is difficult to find high-voltage stress semiconductors or too expensive to apply in conventional boost converters due to limitations of semiconductor technology. In

© Springer International Publishing AG 2018

G. Zhang et al., *Designing Impedance Networks Converters*, Studies in Systems, Decision and Control 119, DOI 10.1007/978-3-319-63655-9_6

practical applications, the voltage gain of a conventional boost converter for a given duty-cycle d of the switching signal, namely $M_{con} = (1 - d)^{-1}$, is limited to about scale of 5-6, which is far away from industrial needs [18]. The flyback converters possess a higher voltage gain than the one of conventional boost converter, but at the expense of large corresponding leakage inductance and a complex structure [19].

To reach specified voltages, N boost converters can be cascaded to reach the voltage gain $M_N = (1 - d)^{-N}$. Cascaded boost converters are, however, too complex due to the additional switches and control units which also reduce system reliability [20]. A high-frequency isolation DC-DC converter with high transformer turn-ratio is applied to solve this problem [21], even though its efficiency is reduced due to its additional stage. This solution also compromises structure, volume and weight.

In industrial applications, quadratic boost converters are the most popular cascaded ones, in which two switches and control circuits are excessively used, but which cannot ensure sufficient voltage gains. Further, many improvements were realized by cascading towards reducing current ripples and to achieve high voltage gains [22–26], e.g. a high-voltage-gain converter based on bootstrap capacitors and boost inductors [27], an optimized cascaded DC-DC converter with a high voltage gain for grid-connected systems [28] and different kinds of interleaved high-step-up converters [29, 30].

Since the insufficient voltage gains in practical applications, power converter with high voltage gains are required especially in solar energy systems. In terms of the impedance networks matching, the disadvantage of traditional boost converter is revealed, and in terms of the proposed methodology, two kinds of 3-Z-network DC-DC boost converters are designed.

6.2 System Design

6.2.1 Disadvantage of Traditional Boost Converters

A traditional boost converter is shown in Fig. 6.1 with the dash diagram indicating the corresponding two-port impedance network.

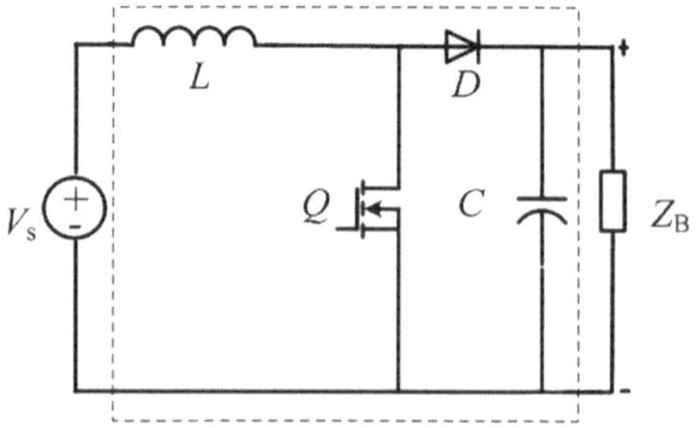

Fig. 6.1 A boost DC-DC converter

From (2.3), one can obtain the transmission parameters matrix of the boost converter, and substituting the parameters into (4.1) results in the input impedance as

$$Z_{\mathrm{Bi}}(j\omega) = j\omega L. \tag{6.1}$$

Similarly, one can also obtain different output impedances in different operational modes. Denote the duty cycle of the boost converter as d. Substituting d into (5.3) results in the average output impedance of the boost converter as

$$Z_{\mathrm{Bo}}(j\omega) = j\left(\frac{\omega^2 LC - d}{(1-\omega^2 LC)\omega C}\right). \tag{6.2}$$

Assume the source impedance of the boost converter in Fig. 6.1 as $Z_{\mathrm{S}}(j\omega) = 0$. In terms of (4.7), when $Z_{\mathrm{Bo}}(j\omega) + Z_{\mathrm{L}}(j\omega) = 0$, one can obtain the largest voltage gain as

$$M_{\max} = \frac{|V_{\mathrm{L}}(j\omega)|}{|V_{\mathrm{S}}(j\omega)|} = \sqrt{1 + \left(\frac{\mathrm{Im}(-Z_{\mathrm{Bo}}(j\omega))}{\mathrm{Re}(Z_{\mathrm{L}}(j\omega))}\right)^2}. \tag{6.3}$$

It is obvious that the larger the value $\mathrm{Im}(Z_{\mathrm{Bo}}(j\omega))$ is, the larger the voltage gain will be. Moreover, in terms of (6.2) the smaller d is, the larger the output impedance and the voltage gain can be obtained. Therefore, impedance networks are cascaded to increase the output impedance so as to realize high voltage gain.

It is thus summarized that the output impedance and the load impedance can not be matched to realize high output voltage in the traditional boost converter. In order to realize higher voltage gains, impedance networks can be cascaded to increase the output impedance for a high voltage gain.

6.2.2 *Selection of Impedance Networks*

The basic cascade type of the two-port impedance network in Table 5.2 is adopted in the traditional boost converter. Replacing the inductor in Fig. 6.1 with a cascaded active two-port impedance network shown in Fig. 6.2 leads to the improved boost converter as shown in Fig. 6.3, where $L_1 = L_2 = L$.

In terms of the voltage-second balance characteristics of inductors, one can obtain

$$\int_0^{dT} (v_{L_1})dt + \int_{dT}^{T} (v_{L_1})dt = \int_0^{dT} V_{\mathrm{S}}dt + \int_{dT}^{T} \frac{V_{\mathrm{S}} - V_{C_1}}{2}dt = 0 \tag{6.4}$$

from Fig. 6.3.

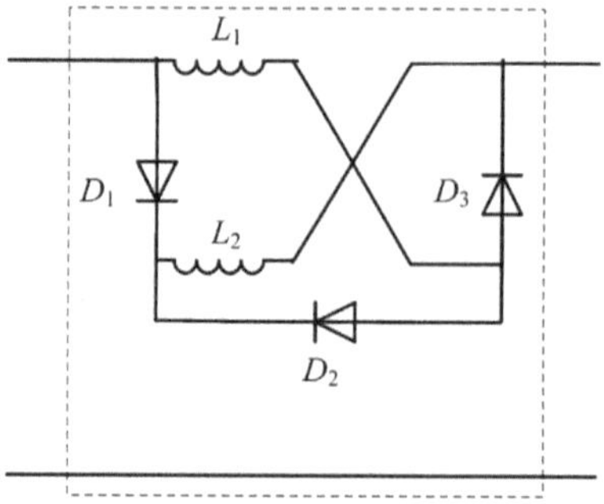

Fig. 6.2 A proposed impedance network

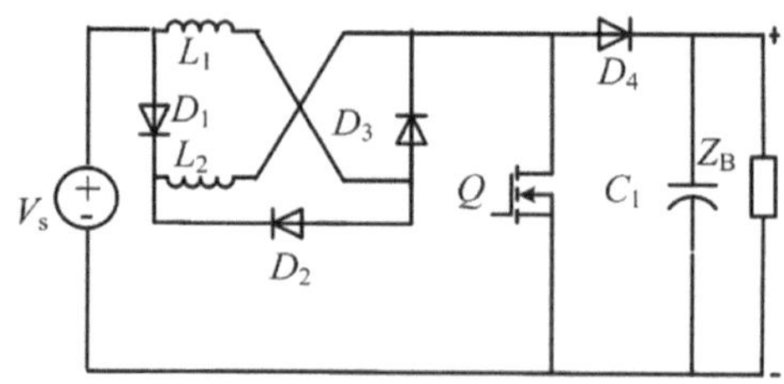

Fig. 6.3 An improved boost converter

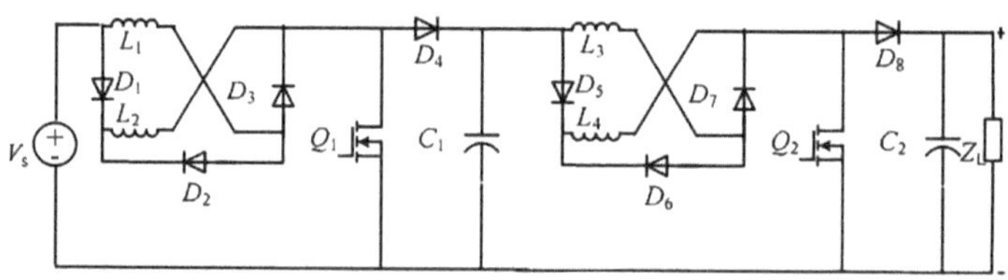

Fig. 6.4 A cascaded boost converter

Solving (6.4) results in the output voltage as

$$V_o = V_{C_1} = V_S \left(\frac{1+d}{1-d} \right). \tag{6.5}$$

It is seen from (6.5) that the output voltage gain is $(1+d)$ times of that of the boost converter, which is, however, not large enough for boosting the voltage in solar energy systems.

In terms of the equations in Table 5.1, one can obtain that the output impedance of two cascaded two-port networks is the multiple of each two-port network, so the cascaded connectivity method can be used to increase the output impedance. Therefore, cascading two identical converters in Fig. 6.3 will give rise to a novel boost converter as shown in Fig. 6.4.

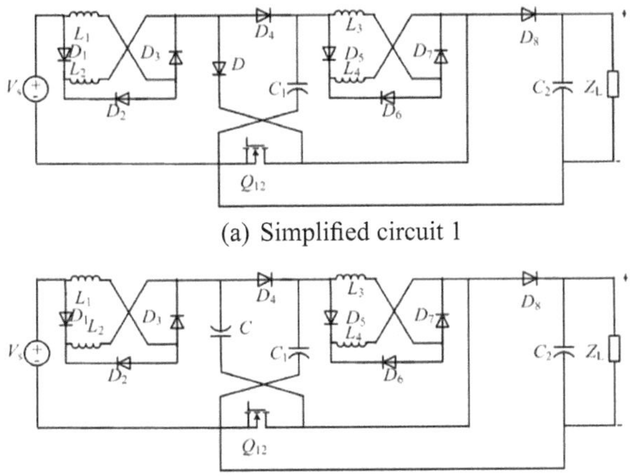

(a) Simplified circuit 1

(b) Simplified circuit 2

Fig. 6.5 Simplified circuits of Fig. 6.4

Such a design leads to some reductant components in the cascaded converters, which need to be simplified using the method proposed in [31], and two simplified converters are depicted in Fig. 6.5. One of the converter is analysed as an example as follows [32, 33].

6.2.3 *Calculation of Input and Output Impedances*

In terms of the input and output impedances of two-port impedance networks in Table 5.3, the input impedance (Z_{Bi}) and output impedance (Z_{Bo}) of the converter in Fig. 6.5a write

$$Z_{\mathrm{Bi}}(j\omega) = \begin{cases} j\frac{1}{2}\omega L, & \text{if } Q_{12} \text{ is on,} \\ j2\omega L + \dfrac{j2\omega L + \dfrac{Z_{\mathrm{B}}(j\omega)}{1 + j\omega Z_{\mathrm{Bi}}(j\omega)C}}{1 + \left(j2\omega L + \dfrac{Z_{\mathrm{B}}(j\omega)}{1 + j\omega Z_{\mathrm{Bi}}(j\omega)C}\right) j\omega C}, & \text{otherwise,} \end{cases} \tag{6.6}$$

and

$$Z_{\mathrm{Bo}}(j\omega) = \begin{cases} -j\dfrac{1}{\omega C}, & \text{if } Q_{12} \text{ is on,} \\ j\dfrac{-4\omega^3 L^2 C + 4\omega L}{4\omega^4 L^2 C^2 - 6\omega^2 LC + 1}, & \text{otherwise.} \end{cases} \tag{6.7}$$

It is remarked that the short-circuited input impedance in (6.6) satisfies the condition in (4.4), i.e. the short-circuited input impedance is inductive, therefore, it realizes input impedance matching. In terms of (6.7), one can obtain the desired voltage gain by adjusting d.

6.3 Operational Modes Analysis

The converter in Fig. 6.5a is taken as an example to be analysed.

The converter contains three active impedance networks in Fig. 6.6, i.e. Z-networks 1, 2 and 3, and is thus named as 3-Z-network boost converter. Therein, Z-network 1 functions as the first boost part, constituted of inductors L_1 and L_2, diodes D_1, D_2 and D_3; Z-network 2 serves as the switch part, composed of switch Q, capacitor C_1, diodes D_4 and D_5; and Z-network 3 acts as the second boost part, consisted of L_3 and L_4, diodes D_6, D_7 and D_8.

It is remarked that a distinct feature of the proposed converter is that there is only one switch used.

The instability of the output voltage in the solar energy leads to different operational modes in the 3-Z-network boost converter, i.e. continue conduction mode (short for CCM) and discontinue conduction mode (short for DCM).

For simplicity, it is assumed that (a) all the components are ideal; (b) the freewheeling diode of the switch is ignored; (c) $L_1 = L_2$ and $L_3 = L_4$.

In the periodic states (on and off) of the switch Q, the inductors charge and discharge alternately. Correspondingly, their currents increase and decrease alternately. Then, there correspond some cases to the discontinuous current states of the inductors, named as discontinuous current cases.

The states of the diodes and the currents of the capacitors correspond to six operational cases, including two CCMs and four DCMs, whose corresponding equivalent circuits are shown in Fig. 6.7a–f. Therein, v_{L_i}, $i = 1, 2, \ldots, 4$ are voltages of L_i, $i = 1, 2, \ldots, 4$, respectively. Assume the clockwise direction as the positive direction of the reference currents, and the arrows in Fig. 6.7a refer to the positive directions of the inductors reference voltages. Moreover, the states of the components in the circuit are detailed in Table 6.1.

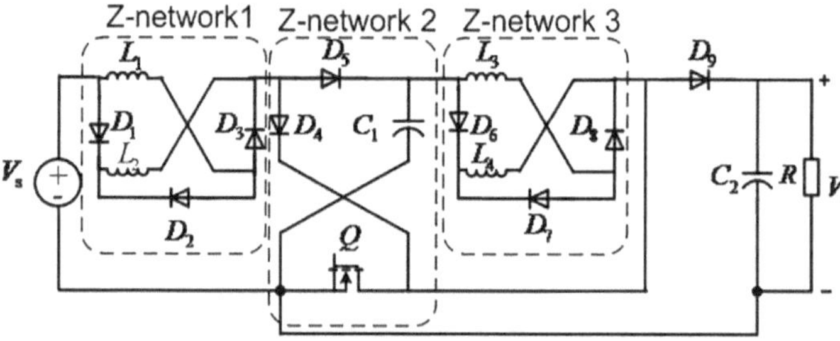

Fig. 6.6 3-Z-network boost converter

Fig. 6.7 Equivalent circuits of 3-Z-network boost converter

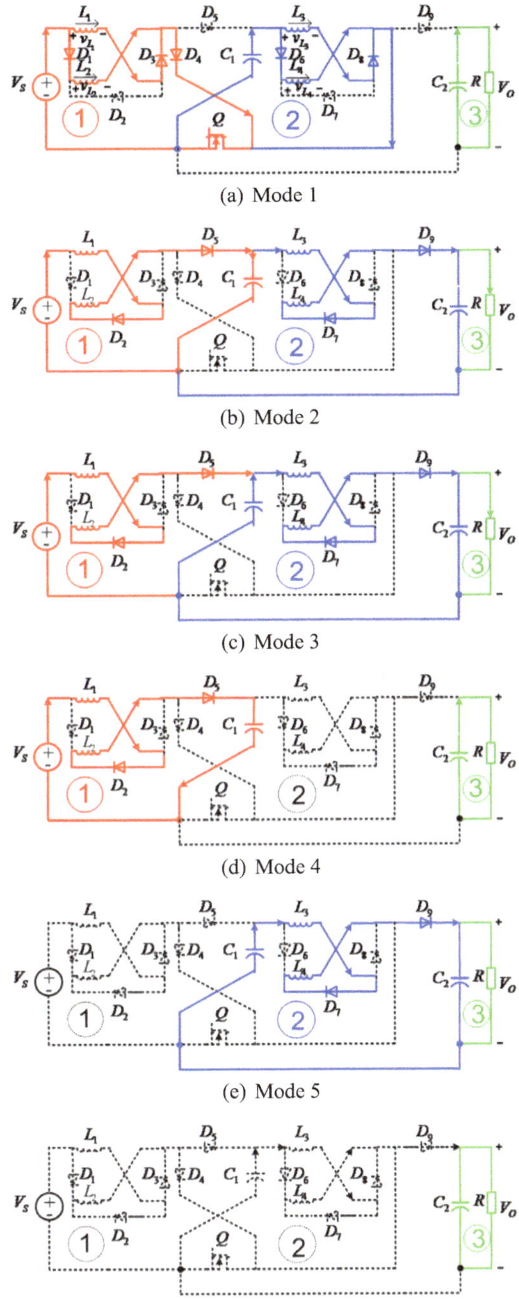

(a) Mode 1

(b) Mode 2

(c) Mode 3

(d) Mode 4

(e) Mode 5

(f) Mode 6

Table 6.1 States of switching components in different modes

	Mode 1	Mode 2	Mode 3	Mode 4	Mode 5	Mode 6
Q	on	off	off	off	off	off
D_1 & D_3	on	off	off	off	off	off
D_2	off	on	on	on	off	off
D_4	on	off	off	off	off	off
D_5	off	on	on	on	off	off
D_6 & D_8	on	off	off	off	off	off
D_7	off	on	on	off	on	off
D_9	off	on	on	off	on	off

6.4 Continuous Conduction Mode

It is known from the analysis above that there are continuous or discontinuous currents of inductors under some combinations of the inductances, the load and the duty. When the currents of inductors in Fig. 6.6 are all continuous, the converter operates in the continuous operational modes. Based on the different current directions of the capacitor C_1, the continuous modes can be classified into two modes, i.e. Case 1: Mode 1 $\rightarrow$ Mode 2; Case 2: Mode 1 $\rightarrow$ Mode 2 $\rightarrow$ Mode 3, whose corresponding modes transition diagrams are shown in Fig. 6.8, where a mode stands for an operational mode.

(1) Case 1

There are two modes in this case, namely, Modes 1 and 2 as shown Fig. 6.8a, whose equivalent circuits are illustrated in Fig. 6.7a and b.

Denote d as the duty of the switch Q, t_0 the initial time of one period, t_1 the mode transition instant from Mode 1 to Mode 2, and $t_2 = T$ the end of the period. In order to describe the operation process of the converter in Case 1, key waveforms of the proposed converter in the steady state are given in Fig. 6.9a, where two modes are marked in two different colors within one period.

Moreover, Fig. 6.9a(1) describes the driven voltage v_g of the switch Q; Fig. 6.9a(2) illustrates the waveform of $i_{L_1}(i_{L_2})$, which is composed of two parts with the blue

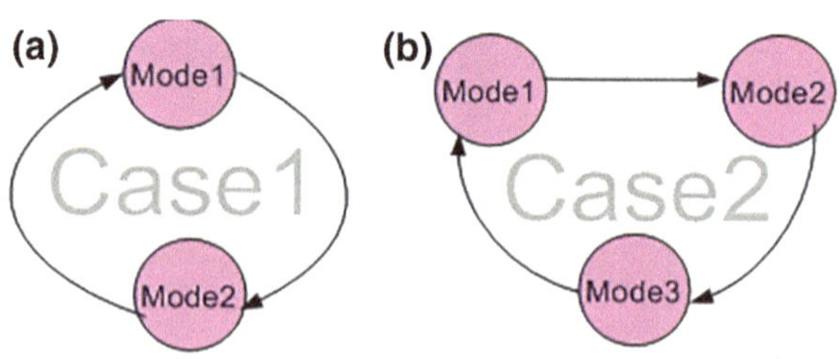

Fig. 6.8 Transitions between modes in CCMs. **a** Case 1; **b** Case 2

Fig. 6.9 Key waveforms of the 3-Z-network boost converter in CCMs

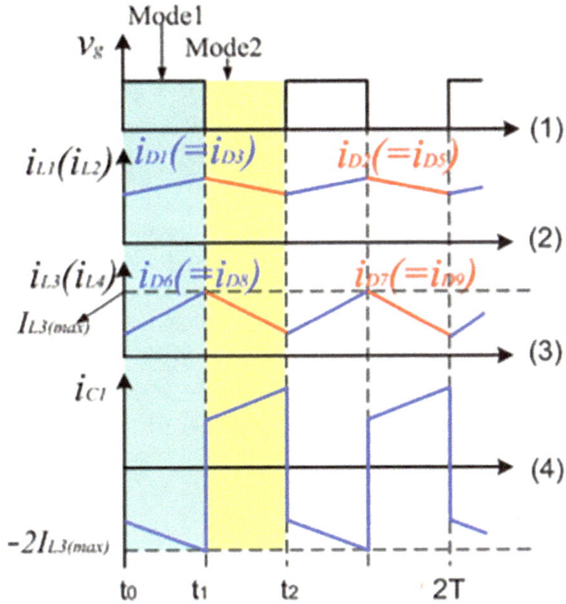

(a) Case 1

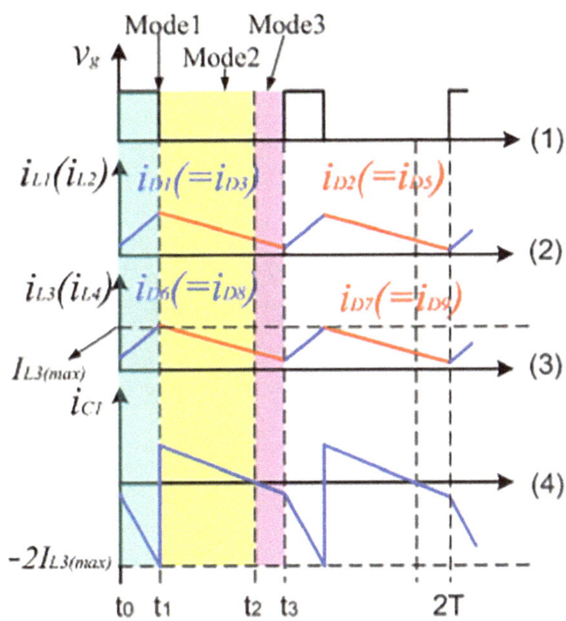

(b) Case 2

one referring to the waveform of $i_{D_1}(= i_{D_3})$ and the red one to the waveforms of $i_{D_2}(= i_{D_5})$; Fig. 6.9a(3) depicts the waveform of $i_{L_3}(i_{L4})$, which is also composed of two parts with the blue one referring to the waveform of $i_{D_6}(= i_{D_8})$ and the red one to the waveform of $i_{D_7}(= i_{D_9})$; Fig. 6.9a(4) portrays the waveforms of i_{C_1}. Therein, $i_{L_i},\ i = 1, 2, \ldots, 4, i_{D_j},\ j = 1, 2, \ldots, 9$, and $i_{C_k},\ k = 1, 2$ are the currents of $L_i,\ i = 1, 2, \ldots, 4,\ D_j,\ j = 1, 2, \ldots, 9$, and $C_k,\ k = 1, 2$, respectively.

Mode 1: $t \in [t_0, t_1]$

As shown in Fig. 6.7a, there are three loops in the circuit, marked in different colors, and the arrows in the circuit refer to the current directions in each loop. As Q turns on, the diodes D_1, D_3 and D_4 undertake positive voltages and turn on synchronously; meanwhile, D_2 bears negative voltage and turns off. Thereafter, L_1 and L_2 are connected in parallel and then cascaded with D_4, Q and V_S to form loop 1 with red lines. The source V_S discharges the energy to L_1 and L_2, then i_{L_1} and i_{L_2} increase, and L_1 and L_2 store the energy. The waveforms (blue lines) of i_{D_1}, i_{D_3}, i_{L_1} and i_{L_2} are shown in Fig. 6.9a(2), where $i_{D_1} = i_{D_3} = i_{L_1} = i_{L_2}$; and i_{D_4}, which is the current of D_4, endures $i_{L_1} + i_{L_2}$, namely, $i_{D_4} = i_{L_1} + i_{L_2} = 2i_{L_1}$.

Accordingly, one has

$$\begin{cases} i_{D_1} = i_{D_3} = i_{L_1} = i_{L_2} \ , \\ i_{D_2} = 0 \ , \\ i_{D_4} = i_{D_1} + i_{D_3} = 2i_{L_1} \ , \\ v_{L_1} = V_S \ , \\ v_{L_2} = V_S \ , \end{cases} \tag{6.8}$$

where v_{L_1} and v_{L_2} are the voltages of L_1 and L_2, respectively.

Meantime, D_5 and D_7 undertake negative voltages and turn off, yet D_6 and D_8 endure positive voltages and turn on. Accordingly, L_3 and L_4 are connected in parallel and then cascaded with Q and C_1 to form loop 2 in blue color. C_1 discharges the energy to L_3 and L_4, and i_{L_3} and i_{L_4} increase. Thus, L_3 and L_4 store energy. The waveforms of i_{D_6}, i_{D_8}, i_{L_3} and i_{L_4} are shown in blue color in Fig. 6.9a(3), and the waveform of i_{C_1} is shown in Fig. 6.9a(4), where $i_{D_6} = i_{D_8} = i_{L_3} = i_{L_4}$ and $i_{C_1} = -2i_{L_3}$, respectively.

Then, one has

$$\begin{cases} i_{D_6} = i_{D_8} = i_{L_3} = i_{L_4} \ , \\ i_{D_5} = 0 \ , \\ i_{C_1} = -2i_{L_3} \ , \\ v_{L_3} = v_{C_1} \ , \\ v_{L_4} = v_{C_1} \ , \end{cases} \tag{6.9}$$

where i_{L_3}, i_{L_4}, v_{L_3}, v_{L_4} and v_{C_1} are the currents of L_3, L_4, and the voltages of L_3, L_4 and C_1, respectively.

Meanwhile, D_9 endures the negative voltage and turns off, then the capacitor C_2 and the load R are cascaded to form loop 3 in green color. Therein, C_2 discharges the energy to R, then the output voltage of the converter v_o reads

$$v_o = v_{C_2}, \tag{6.10}$$

where v_{C_2} is the voltage of capacitor C_2.

Mode 2: $t \in [t_1, t_2]$

At t_1, Q turns off, the mode changes from Mode 1 to Mode 2, as shown in Fig. 6.7b.

As Q is off, D_1, D_3, D_4, D_6 and D_8 undertake negative voltage and turn off, yet D_2, D_5, D_7 and D_9 turn on and then form three loops in this mode. Therein, loop 1 is marked with red color, namely, V_S-L_1-D_2-L_2-D_5-C_1, where V_S, L_1 and L_2 discharge energy to C_1, namely $V_S = v_{L_1} + v_{L_2} + v_{C_1}$. Moreover, i_{L_1} and i_{L_2} decrease as the red lines shown in Fig. 6.9a(2), and the currents of D_2 and D_5 are equal to i_{L_1} for the cascaded connectivity, and i_{C_1} increases as shown in Fig. 6.9a(4), namely

$$\begin{cases} i_{D_2} = i_{D_5} = i_{L_1} = i_{L_2} , \\ i_{D_1} = i_{D_3} = i_{D_4} = 0 , \\ v_{L_1} + v_{L_2} = V_S - v_{C_1} . \end{cases} \tag{6.11}$$

V_S, L_1, D_2, L_2, D_5, L_3, D_7, L_4, D_9 and C_2 form loop 2 marked with red and blue lines, where V_S, L_1, L_2, L_3 and L_4 discharge the energy to C_2 and R, namely, $V_S = v_{L_1} + v_{L_2} + v_{L_3} + v_{L_4} + v_{C_2}$, and i_{C_2} decreases due to the discharge energy to the load R. Moreover, i_{L_3} and i_{L_4} decrease as the red lines shown in Fig. 6.9a(3), and the currents of D_7 and D_9 are equal to i_{L_3} for the cascaded connectivity, namely

$$\begin{cases} i_{D_7} = i_{D_9} = i_{L_3} = i_{L_4} , \\ i_{D_6} = i_{D_8} = 0 , \\ v_{L_3} + v_{L_4} = V_S - (v_{L_1} + v_{L_2} + v_{C_2}) . \end{cases} \tag{6.12}$$

(2) Case 2

Case 1 and 2 are both CCMs. Nevertheless, different to Case 1, there are three modes in Case 2 due to the direction of i_{C_1}, i.e. Modes 1, 2 and 3, whose equivalent circuits are shown in Fig. 6.7a, b and c.

Denote t_0 the beginning of one period, t_1 the mode transition instant from Mode 1 to Mode 2, i.e. $t_1 = t_0 + dT$, t_2 the mode transition instant from Mode 2 to Mode 3, and $t_3 = T$ the end of the period. In order to describe the operation process of the converter, key waveforms of the proposed converter in Case 2 are shown in Fig. 6.9b. Therein, three modes are marked with three different colors in one period.

The operation process of the proposed converter in a switch period is analysed in the following section according to the waveforms in Fig. 6.9b.

Mode 1: $t \in [t_0, t_1]$

The process is similar as the descriptions of Mode 1 of Case 1.

Mode 2: $t \in [t_1, t_2]$

The process is similar to the descriptions of Mode 2 of Case 1 except i_{C_1}. Therein, i_{C_1} decreases as shown in Fig. 6.9b(4), which is different from the increase of i_{C_1} in Fig. 6.9a(4), because the energy stored in the inductors are not enough to charge the load in this case.

Mode 3: $t \in [t_2, t_3]$
At t_2, i_{C_1} and i_{C_2} decrease to 0, then Mode 3 appears, shown as Fig. 6.7c, where Q, D_1, D_3, D_4, D_6, D_7, D_8 and D_9 are off, D_2 and D_5 are on. Thus, there are also two loops marked in different colors.

Loop 1 is the same as loop 1 of Mode 2.

In loop 2, i_{C_1} decreases from 0 to negative, which means that not only V_S, L_1 and L_2, but also C_1 charge the energy to the following circuit, and they fulfill the same equations as in Case 1.

6.5 Discontinuous Conduction Mode

In terms of the inductor currents i_{L_1}, i_{L_2}, i_{L_3}, i_{L_4} and the direction of the capacitor current i_{C_1}, there are four DCMs in the proposed 3-Z-network boost converter, i.e. Case 3: Mode 1 → Mode 2 → Mode 4; Case 4: Mode 1 → Mode 2 → Mode 3 → Mode 5; Case 5: Mode 1 → Mode 2 → Mode 4 → Mode 6; Case 6: Mode 1 → Mode 2 → Mode 3 → Mode 5 → Mode 6, as shown in Fig. 6.10, and the corresponding states of the components are shown in Table 6.1.

(3) Case 3
Different from Case 1 and 2, Case 3 is a DCM when i_{L_3} and i_{L_4} are discontinuous. There are three modes in this case, i.e. Modes 1, 2 and 4, whose equivalent circuits are shown in Fig. 6.7a, b and d.

Denote t_0 the beginning of one period, t_1 the mode transition instant from Mode 1 to Mode 2, i.e. $t_1 = t_0 + dT$, t_2 the mode transition instant from Mode 2 to Mode 4, and $t_3 = T$ the end of the period. The key waveforms of the proposed converter in Case 3 shown in Fig. 6.11a can describe the operation process, where three modes are marked in three different colors in one period.

The operation process of the proposed converter in a switch period is analysed as follows in terms of the waveforms in Fig. 6.11a.

Mode 1: $t \in [t_0, t_1]$
The process is similar as that one in Mode 1 in Case 1.

Mode 2: $t \in [t_1, t_2]$
The process is similar as that one in Mode 2 in Case 1.

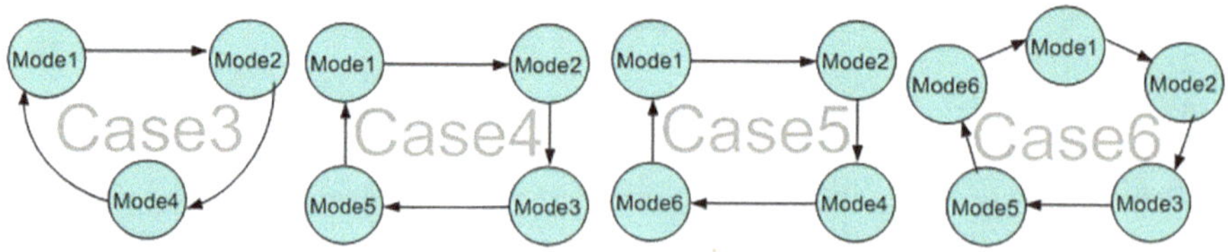

Fig. 6.10 Transitions between modes in DCMs. **a** Case 3. **b** Case 4. **c** Case 5. **d** Case 6

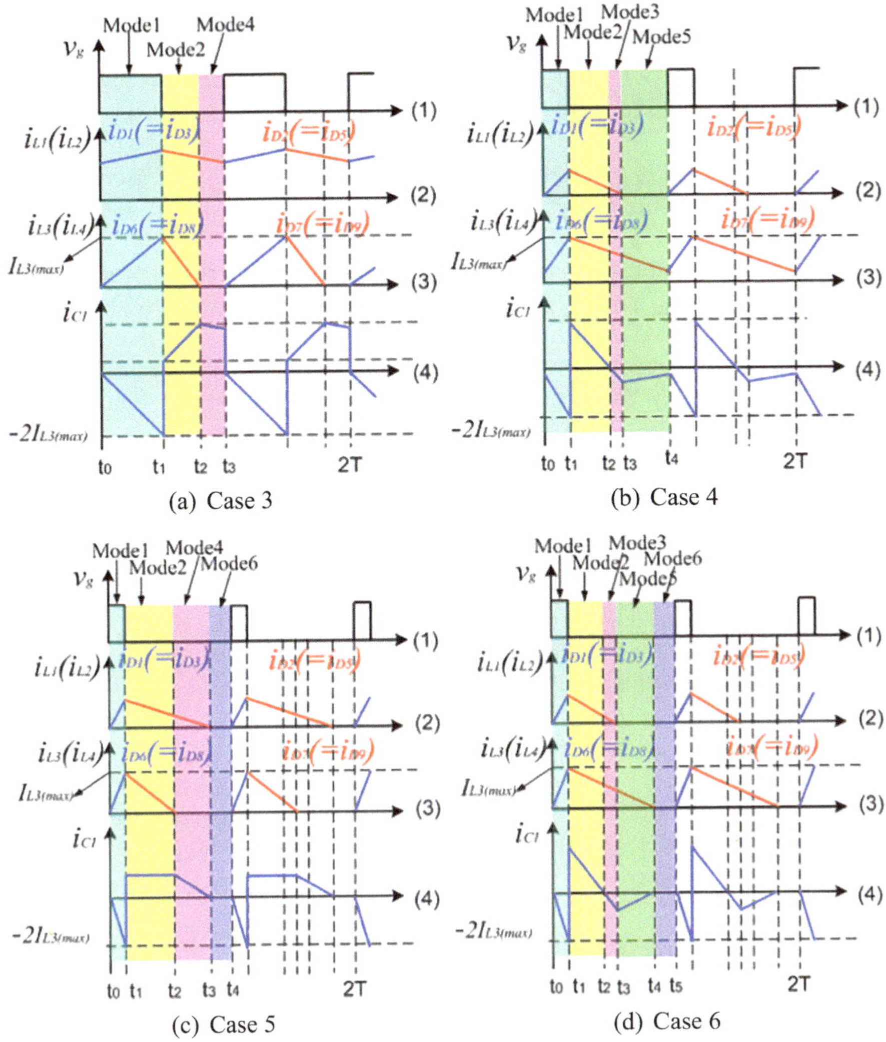

Fig. 6.11 Key waveforms of 3-Z-network boost converter in DCMs

Mode 4: $t \in [t_2, t_3]$

When i_{L_3} and i_{L_4} decrease to 0 and Q keeps off, Mode 4 exists, which is described by $i_{D_7} = i_{D_9} = i_{L_3} = i_{L_4} = 0$. Then D_7 and D_9 are off, the waveforms are shown in Fig. 6.11a. As Fig. 6.7c shows, Q, D_1, D_3, D_4, D_6, D_7, D_8 and D_9 are off, but D_2 and D_5 are on, there are also two loops marked in different colors.

From loop 1 marked in red color, i.e. V_S-L_1-D_2-L_2-D_5-C_1, one has $i_{C_1} = i_{D_2} = i_{D_5} = i_{L_1} = i_{L_2}$. The source V_S, L_1 and L_2 discharge the energy to C_1, so that v_{C_1} increases, namely

$$\begin{cases} i_{C_1} = i_{D_2} = i_{D_5} = i_{L_1} = i_{L_2} \,, \\ i_{D_1} = i_{D_3} = i_{D_4} = 0 \,. \end{cases} \tag{6.13}$$

In loop 2 marked in blue color, C_2 discharges energy to R.

(4) Case 4

Case 4 is also a DCM. It corresponds to the case when i_{L_1} and i_{L_2} are discontinuous. There are four modes in this case, i.e. Modes 1, 2, 3, and 5, and their equivalent circuits are shown in Fig. 6.7a, b, c and e.

Denote t_0 the beginning of one period, t_1 the mode transition instant from Mode 1 to Mode 2, i.e. $t_1 = t_0 + dT$, t_2 the mode transition instant from Mode 2 to Mode 3, t_3 the mode transition instant from Mode 3 to Mode 5, and $t_4 = T$ the end of the period. Key waveforms of the proposed converter in Case 4 as shown in Fig. 6.11b illustrate the operation process. Therein, four modes are marked in four different colors in one period.

The operation process in a switch period is analysed in the following content according to the waveforms in Fig. 6.11b.

Mode 1: $t \in [t_0, t_1]$

The process is same as the one of Mode 1 in Case 2.

Mode 2: $t \in [t_1, t_2]$

The process is same as the one of Mode 2 in Case 2.

Mode 3: $t \in [t_2, t_3]$

The process is also similar as that one in Mode 2 in Case 2. However, i_{L_1} and i_{L_2} decrease to be 0 shown in Fig. 6.11b(2), which is different from the waveforms of i_{L_1} in Fig. 6.9b(2). Thereafter, Mode 5 occurs.

Mode 5: $t \in [t_3, t_4]$

When i_{L_1} and i_{L_2} decrease to be 0, loop 1 is off. Then, C_1, L_3 and L_4 discharge the energy to C_2 and R. Meanwhile, C_2 also discharge the energy to R. Hence, i_{C_2} decrease from positive to negative. Thereafter, C_2, C_1, L_3 and L_4 discharge the energy to R.

(5) Case 5

When i_{L_1} and i_{L_2} decrease to 0 in Mode 4 of Case 3, there is only loop 3 left, as shown in Fig. 6.7f. It is Case 5. There are four modes in this case, i.e. Modes 1, 2, 4, and 6, whose equivalent circuits are shown in Fig. 6.7a, b, d, and f.

Denote t_0 the beginning of one period, t_1 the mode transition instant from Mode 1 to Mode 2, i.e. $t_1 = t_0 + DT$, t_2 the mode transition instant from Mode 2 to Mode 4, t_3 the mode transition instant from Mode 4 to Mode 6, and $t_4 = T$ the end of the period. Key waveforms of the proposed converter in Case 5 as shown in Fig. 6.11c illustrate the operation process. Therein, four modes are marked in four different colors in one period. The operational process in a switch period is analysed in the following content according to the waveforms in Fig. 6.11c.

Mode 1: $t \in [t_0, t_1]$

The process is same as the one of Mode 1 in Case 3.

Mode 2: $t \in [t_1, t_2]$
The process is same as the one of Mode 2 in Case 3.

Mode 4: $t \in [t_2, t_3]$
The process is same as the one of Mode 3 in Case 3.

Mode 6: $t \in [t_3, t_4]$
i_{L_1} and i_{L_2} decrease to 0 and Q keeps off, then Mode 6 occurs, as shown in Fig. 6.7f. Therein, C_2 discharges the energy to R, and i_{C_2} is constant.

(6) Case 6
When i_{L_3} and i_{L_4} decrease to 0 in Mode 5 of Case 4, then there is only loop 3 left shown in Fig. 6.7f, which is Case 6. There are five modes in this case, i.e. Mode 1, Mode 2, Mode 3, Mode 5, and Mode 6, whose equivalent circuits are shown in Fig. 6.7a, b, c, e and f.

Denote t_0 the beginning of one period, t_1 the mode transition instant from Mode 1 to Mode 2, i.e. $t_1 = t_0 + dT$, t_2 the mode transition instant from Mode 2 to Mode 3, t_3 the mode transition instant from Mode 3 to Mode 5, t_4 the mode transition instant from Mode 5 to Mode 6, and $t_5 = T$ the end of the period. Key waveforms of the proposed converter in Case 6 as shown in Fig. 6.11d illustrate the operational process. Therein, five modes are marked in five different colors in one period. The operation process in a switch period is analysed in the following according to the waveforms in Fig. 6.11d.

Mode 1: $t \in [t_0, t_1]$
The process is same as the one of Mode 1 in Case 4.

Mode 2: $t \in [t_1, t_2]$
The process is same as the one of Mode 2 in Case 4.

Mode 3: $t \in [t_2, t_3]$
The process is same as the one of Mode 3 in Case 4.

Mode 5: $t \in [t_3, t_4]$
The process is same as the one of Mode 5 in Case 4.

Mode 6: $t \in [t_4, t_5]$
i_{C_1}, i_{L_3} and i_{L_4} decrease to 0 and Q keeps off, then Mode 6 occurs, which is shown in Fig. 6.7f. Therein, C_2 discharges the energy to R, and i_{C_2} is constant.

6.6 Parameters Determination

Normally, the parameters determination of a converter is to determine the rated voltages and rated currents of the components in the circuit.

In order to simplify the analysis, the output voltage V_o is deduced firstly. Because the voltage and current stresses of the components in CCMs are largest, the parameters of the proposed converter in CCM will be chosen as the parameters of the converter.

According to the analyses of Cases 1 and 2, the output voltage V_o is deduced as follows. In terms of the voltage-second balance theory, one has

$$\begin{cases} \int_0^{dT} (v_{L_1} + v_{L_2})dt + \int_{dT}^{T} (v_{L_1} + v_{L_2})dt = 0 \,, \\ \int_0^{dT} (v_{L_3} + v_{L_4})dt + \int_{dT}^{T} (v_{L_3} + v_{L_4})dt = 0 \,. \end{cases} \tag{6.14}$$

Substituting (6.8)–(6.12) into (6.14) leads to

$$\begin{cases} 2V_S dT + (V_S - V_{C_1})(1-d)T = 0, \\ 2V_{C_1} dT + (V_{C_1} - V_o)(1-d)T = 0. \end{cases} \tag{6.15}$$

From (6.15), one has

$$V_o = V_{C_2} = V_S \left(\frac{1+d}{1-d} \right)^2 , \tag{6.16}$$

and

$$V_{C_1} = V_S \frac{1+d}{1-d}. \tag{6.17}$$

According to (6.16), the relationships of duty-cycle d and voltage gain $M = V_o / V_S$ in conventional boost converters, quadratic boost converters, and 3-Z-network boost converters are depicted in Fig. 6.12 in different colors, respectively, with a zoom-in for $0 < d < 0.5$. It is remarked that the proposed converter can reach a much higher voltage-gain than the quadratic and the conventional ones.

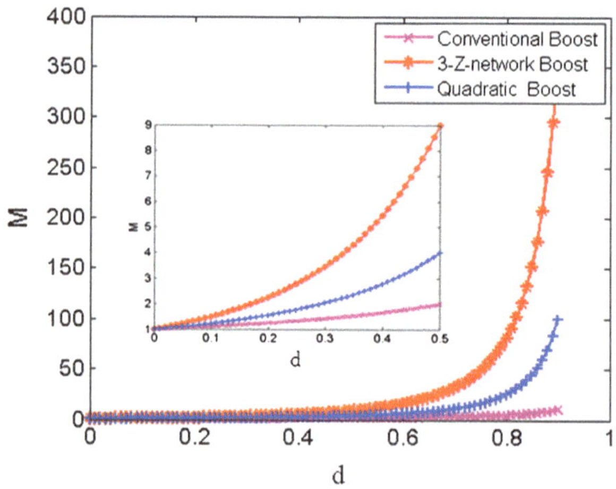

Fig. 6.12 Relationship figure of duty-cycle d and voltage gain M

6.6.1 Output Voltage and Voltage Stress of Electrical Components

In terms of (6.16) and (6.17), the voltages of $L_i,\ i = 1, 2, \ldots, 4$ can be obtained as follows, therein, $V_{L_i},\ i = 1, 2, \ldots, 4$ refer to the average voltages of $L_i,\ i = 1, 2, \ldots, 4$ when Q is on and off, respectively.

$$V_{L_1} = V_{L_2} = \begin{cases} V_S, & \text{if Q is on,} \\ -V_S \dfrac{d}{1-d}, & \text{otherwise,} \end{cases} \tag{6.18}$$

and

$$V_{L_3} = V_{L_4} = \begin{cases} V_S \dfrac{1+d}{1-d}, & \text{if Q is on,} \\ -V_S \dfrac{d(1+d)}{(1-d)^2}, & \text{otherwise.} \end{cases} \tag{6.19}$$

In terms of the Kirhhoff's Voltage Law (KVL), one can obtain the voltage of each component in the circuit as shown in Table 6.2.

Assuming a lossless circuit leads to $V_S \cdot I_{in} = V_o \cdot I_o$, where I_{in} and I_o are input and output currents, respectively. In terms of (6.16), one has

$$I_{in} = I_o \frac{(1+d)^2}{(1-d)^2}. \tag{6.20}$$

Table 6.2 Voltage stress of each switching component in CCMs

	Mode 1	Mode 2 or Mode 3
v_Q	0	$V_S \frac{(1+d)^2}{(1-d)^2}$
v_{D_1} & v_{D_3}	0	$V_S \frac{d}{1-d}$
v_{D_2}	V_S	0
v_{D_4}	0	$V_S \frac{2d(1+d)}{(1-d)^2}$
v_{D_5}	$V_S \frac{1+d}{1-d}$	0
v_{D_6} & v_{D_8}	0	$V_S \frac{d(1+d)}{(1-d)^2}$
v_{D_7}	$V_S \frac{1+d}{1-d}$	0
v_{D_9}	$V_S \frac{(1+d)^2}{(1-d)^2}$	0

Moreover, I_{in} can also be expressed as

$$I_{\text{in}} = \frac{\int_0^{dT} 2i_{L_1} dt + \int_{dT}^{T} i_{L_1} dt}{T} = (1+d)I_{L_1}. \tag{6.21}$$

Substituting (6.21) into (6.20) results in

$$I_{L_1} = I_{L_2} = I_o \frac{1+d}{(1-d)^2}. \tag{6.22}$$

Similarly, I_{L_3} and I_{L_4} are expressed as

$$I_{L_3} = I_{L_4} = I_o \frac{1}{1-d}. \tag{6.23}$$

In terms of (6.8), (6.9), (6.11), (6.12) and Fig. 6.9a, the average currents of the diodes are written as

$$\begin{cases} I_{D_1} = I_{D_3} = dI_{L_1} = I_o \dfrac{d(1+d)}{(1-d)^2}, \\ I_{D_2} = I_{D_5} = (1-d)I_{L_1} = I_o \dfrac{1+d}{1-d}, \\ I_{D_4} = 2dI_{L_1} = I_o \dfrac{2d(1+d)}{(1-d)^2}, \\ I_{D_6} = I_{D_8} = dI_{L_3} = I_o \dfrac{d}{1-d}, \\ I_{D_7} = I_{D_9} = (1-d)I_{L_3} = I_o, \\ I_Q = d(2I_{L_1} + 2I_{L_3}) = I_o \dfrac{4d}{(1-d)^2}. \end{cases} \tag{6.24}$$

Normally, one derives the voltage of the switches in terms of Table 6.2, while determines the corresponding current in terms of (6.24). Then, substituting the related parameters into the corresponding equations in Table 6.2 and (6.24), one can obtain their largest values and consider their corresponding safety margin to decide their rated values.

6.6.2 *Parameters of Inductors*

The parameter determination of the inductor is to determine the rated current and inductance, given a permitted fluctuation range $x_L\%$ (x_L is pre-assigned), an output voltage V_o, an output current I_o, and a switching period T.

6.6.2.1 Determination of the Rated Current

The rated currents of inductors can be obtained from (6.22) and (6.23).

6.6.2.2 Determination of the Rated Inductance

The ripples of the inductors have also great influence on the stability of the converter; consequently, the inductors must be designed in terms of the permitted ripples.

The inductors in the converter can be determinated based on the differential equation of inductors

$$L = \frac{V_L dt_L}{di_L}, \tag{6.25}$$

where V_L is the voltage of the corresponding inductor as Q is on, $dt_L = d \cdot T$ is the time interval of Q as it is on, and di_L is the current ripple of the corresponding inductor within the time interval dt_L.

Denote the permitted error of I_L by di_L. di_L is restrained by the permitted fluctuation range $x_L\%$ as

$$di_L = x_L\% I_L. \tag{6.26}$$

Substituting (6.26) into (6.25) leads to the inductance of L,

$$L = \frac{V_L dT}{x_L\% I_L}, \tag{6.27}$$

thereafter, substituting the average voltage of inductors in (6.18) and (6.19) for Q on, as well (6.22) and (6.23) into (6.27) leads to

$$\begin{cases} L_1 = L_2 = \dfrac{V_S dT}{x_L\% I_{L_1}} = \dfrac{V_S d(1-d)^2 T}{x_L\% I_o (1+d)}, \\ L_3 = L_4 = \dfrac{V_S (1+d) dT}{(1-d) x_L\% I_{L_3}} = \dfrac{V_S (1+d) dT}{x_L\% I_o}. \end{cases} \tag{6.28}$$

Hence, the range of the inductance can be determined based on $0 \le d \le 1$, and the maximum is taken as the rated inductance.

6.6.3 *Parameters of Capacitors*

Similar to the parameter determination of the inductor, the one of the capacitor is to determine its rated voltage and capacitance, given a permitted fluctuation range, $x_C\%$ (x_C is pre-assigned), an output voltage V_o, an output current I_o, and a switching period T.

6.6.3.1 Determination of the Rated Voltage

The rated voltages of capacitors are determined in terms of (6.16) and (6.17).

6.6.3.2 Determination of the Rated Capacitance

The ripples of the capacitors have great influence on the stability of the converter, whose permitted fluctuation range can be used to determine the capacitance.

Then, the capacitors in converter is determined based on the differential equation of capacitors

$$C = \frac{I_C dt_C}{dv_C}, \tag{6.29}$$

where I_C is the current of the corresponding capacitor as Q is on, $dt_C = dT$ is time interval of Q as it is on , and dv_C is the voltage ripple of the corresponding capacitor within the time interval, dt_C.

Denote the permitted error of V_C by dv_C. dv_C is determined by the permitted fluctuation range $x_C\%$ as

$$dv_C = x_C\% V_C\ . \tag{6.30}$$

When Q is on, $I_{C_1} = 2I_{L_3}$ and $I_{C_2} = I_o$, then substituting these two equations and (6.30) into (6.29) leads to

$$\begin{cases} C_1 = \dfrac{2I_{L_3}dT}{x_C\% V_{C_1}} = \dfrac{2I_o dT}{x_C\% V_s(1+d)}, \\ C_2 = \dfrac{I_o dT}{x_C\% V_o} = \dfrac{I_o(1-d)^2 dT}{x_C\%(1+d)^2 V_s}. \end{cases} \tag{6.31}$$

Therein, the range of the capacitance can be calculated according to (6.31), and the maximal one is taken as the rated capacitance.

6.7 Simulations and Experiments

To verify the feasibility and validity of the proposed converter, PSIM software is applied for the simulation.

The pre-assigned parameters are as follows, $V_S = 12$ V, $V_o = 108$ V, $f =$ 100 kHz and $R = 400\ \Omega$. Further, assume the parameters of the converter as: $C_1 = 220\ \mu$F, $C_2 = 470\ \mu$F, $L_1 = L_2 = 100\ \mu$H, $L_3 = L_4 = 200\ \mu$H. Setting up all the parameters and $d = 0.5$ in the simulation circuit of PSIM, one can obtain the simulation results of Case 1 as shown in Fig. 6.13a.

Fig. 6.13 Simulation results of the 3-Z-network boost converter in CCMs

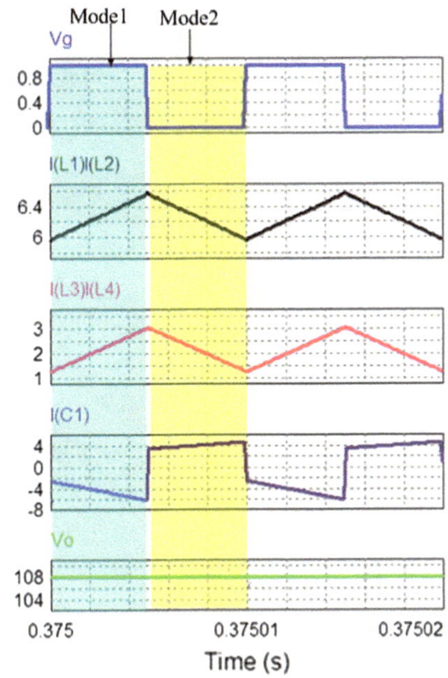

(a) Case 1

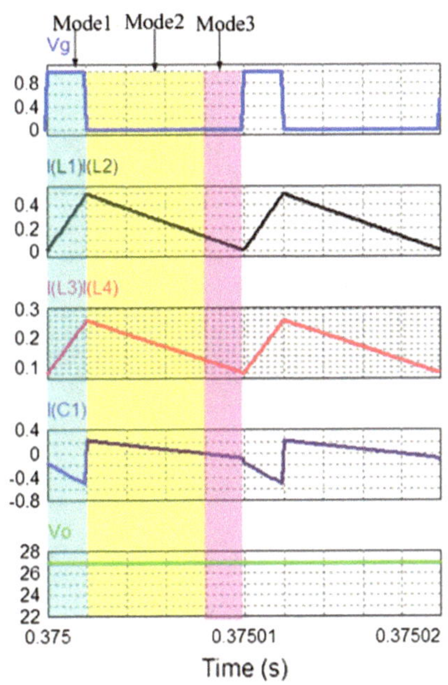

(b) Case 2

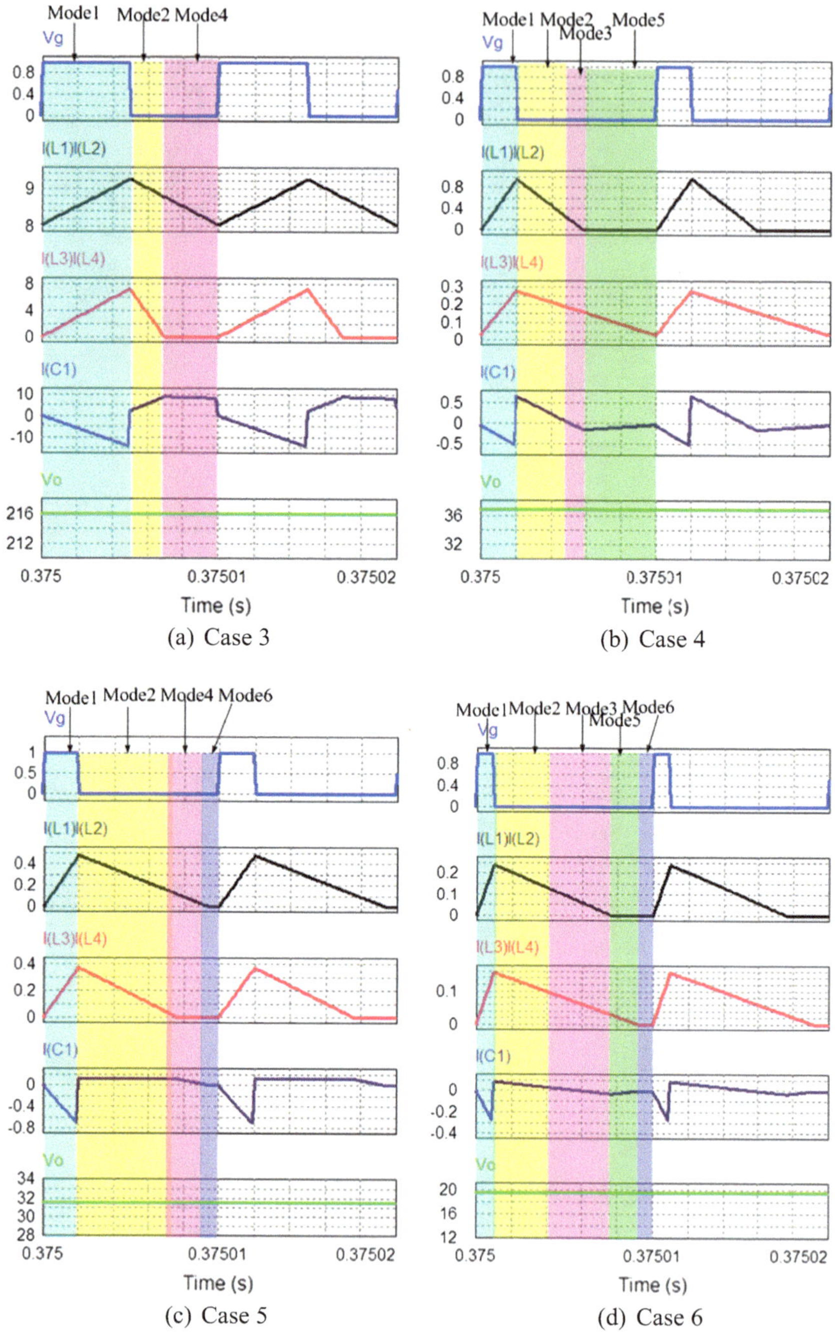

Fig. 6.14 Simulation results of the 3-Z-network Boost converter in DCMs

Then, inductances of L_i, $i = 1, 2, \ldots, 4$, R, and d are reduced so that the energy stored in the inductors is insufficient for supplying to the load, which results in other operational modes with the corresponding parameters shown as follows:

Case 2: $L_1 = L_2 = 50\ \mu\text{H}$, $L_3 = L_4 = 200\ \mu\text{H}$, $R = 200\ \Omega$ and $d = 0.2$;
Case 3: $L_1 = L_2 = 50\ \mu\text{H}$, $L_3 = L_4 = 25\ \mu\text{H}$, $R = 300\ \Omega$ and $d = 0.5$;
Case 4: $L_1 = L_2 = 25\ \mu\text{H}$, $L_3 = L_4 = 200\ \mu\text{H}$, $R = 300\ \Omega$ and $d = 0.2$;
Case 5: $L_1 = L_2 = 50\ \mu\text{H}$, $L_3 = L_4 = 100\ \mu\text{H}$, $R = 300\ \Omega$ and $d = 0.2$; and
Case 6: $L_1 = L_2 = 50\ \mu\text{H}$, $L_3 = L_4 = 100\ \mu\text{H}$, $R = 300\ \Omega$ and $d = 0.1$;

which are shown in Figs. 6.13b and 6.14a–d, respectively.

The simulation results are shown in Figs. 6.13 and 6.14. In the six sub-figures for six cases, from top to bottom are the driven voltage v_g, the current of the inductors $L_1(L_2)$, $L_3(L_4)$ and the capacitance C_1, and the output voltage v_o, respectively.

A prototype of 3-Z-network boost converter is built as shown in Fig. 6.15, where the converter is composed of L_1, L_2, C_1, C_2, switch Q (Type:IRFP250N), D_1-D_8 (Type:SB5100), and D_9 (Type:MBRF20200). In addition, IXDN404 IC is used to drive the switch Q.

Assume the parameters as follows for Case 1: $C_1 = 220\ \mu\text{F}$, $C_2 = 470\ \mu\text{F}$, $L_1 = L_2 = 100\ \mu\text{H}$, $L_3 = L_4 = 200\ \mu\text{H}$, $R = 400\ \Omega$, and $f = 100$ kHz. The waveforms of the converter with an input voltage 12 V are shown in Fig. 6.16, where, from top to bottom, are the Gate-Source voltage of the switch Q, v_{GS}; the Drain-Source voltage of the switch Q, v_{DS}; the current of L_1, i_{L_1}; the current of L_3, i_{L_3}; and the output voltage, v_o. Moreover, the output voltage is about 104.5 V, which is 3.5 V lower than the theoretical value 108 V, and it is reasonable for the diodes voltage

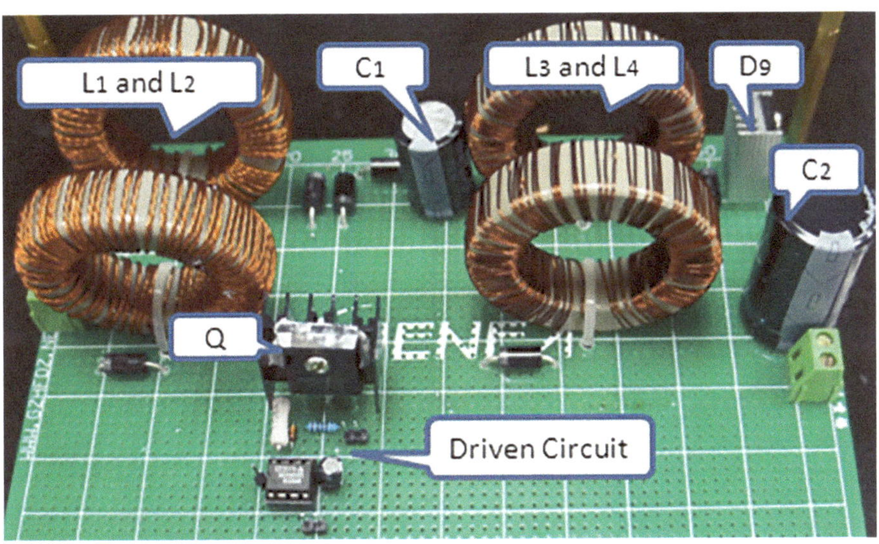

Fig. 6.15 Prototype of the proposed converter

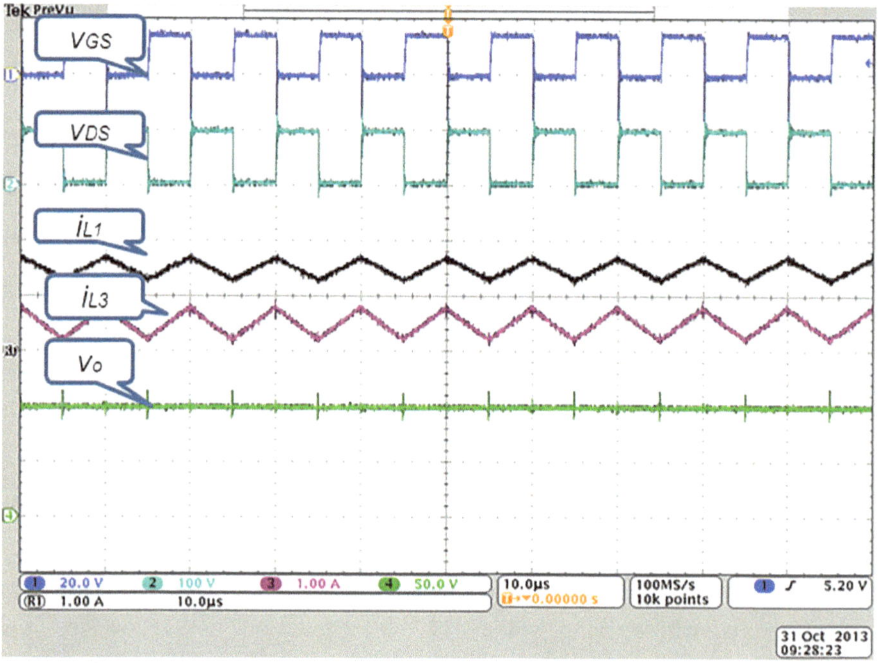

Fig. 6.16 Experimental waveforms of Case 1

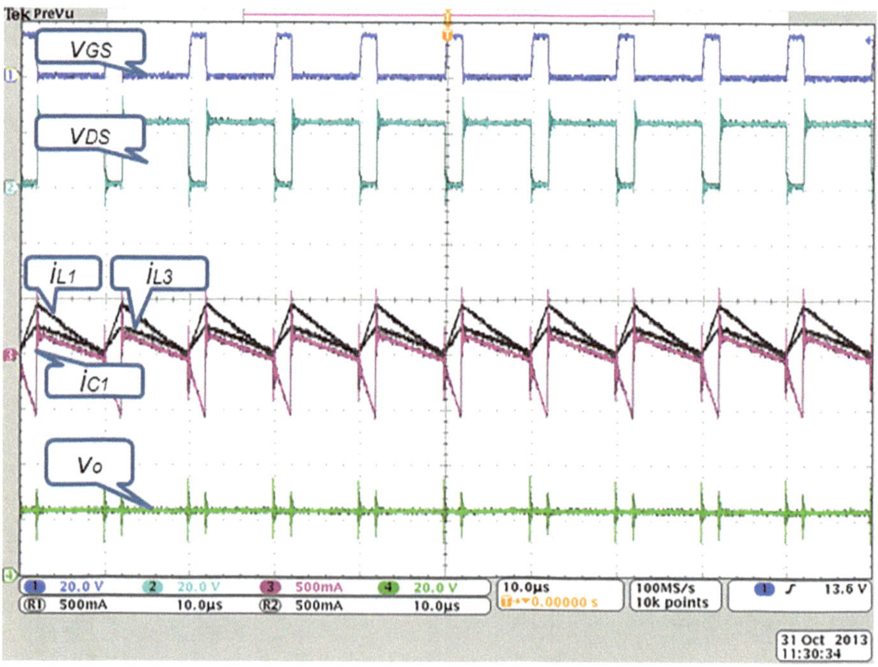

Fig. 6.17 Experimental waveforms of Case 2

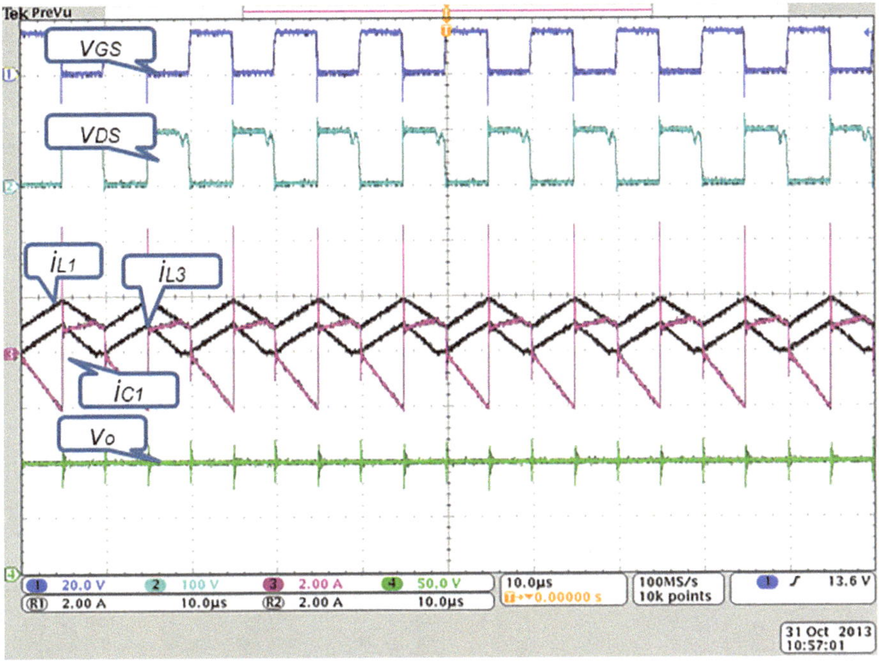

Fig. 6.18 Experimental waveforms of Case 3

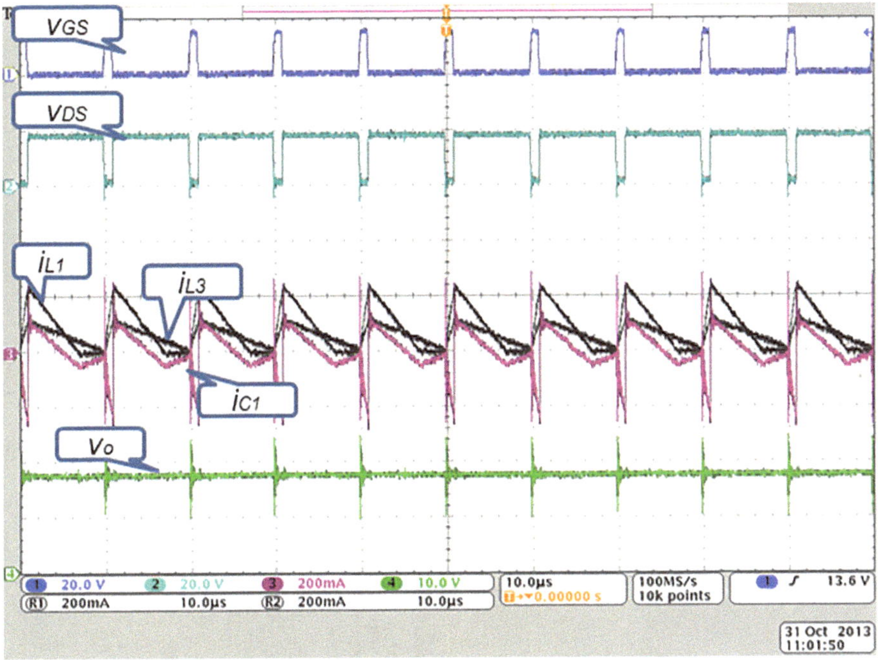

Fig. 6.19 Experimental waveforms of Case 4

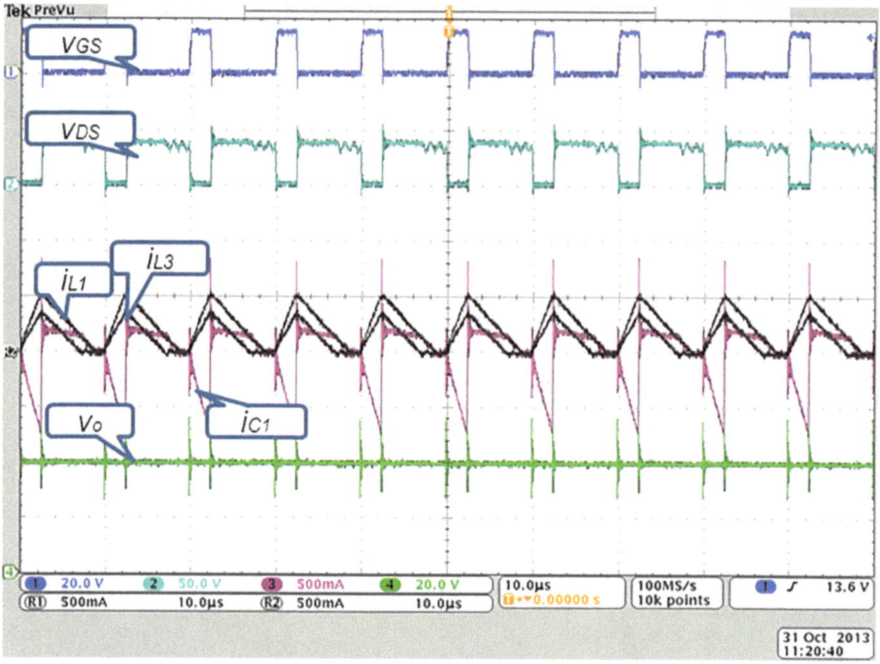

Fig. 6.20 Experimental waveforms of Case 5

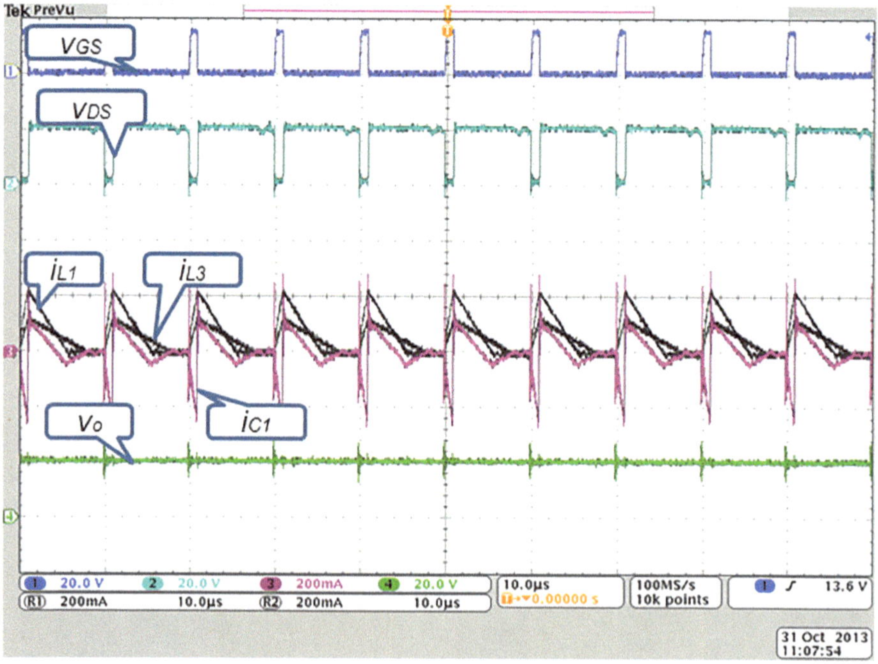

Fig. 6.21 Experimental waveforms of Case 6

drops. Furthermore, the mean currents of i_{L_1} and i_{L_3} are about 1.64 A and 0.53 A, respectively. Hence, the experimental results verify the analytical and simulation results.

Fix the inductors parameters to be $L_1 = L_2 = 45\ \mu$H, $L_3 = L_4 = 140\ \mu$H. Then, tuning the load and the duty results in different cases, i.e. Cases 2–6 as shown in Figs. 6.17, 6.18, 6.19, 6.20 and 6.21, respectively; where the corresponding parameters settings are: Case 2: $d = 0.2$, $R = 200\ \Omega$; Case 3: $d = 0.5$, $R = 500\ \Omega$; Case 4: $d = 0.1$, $R = 300\ \Omega$; Case 5: $d = 0.25$, $R = 500\ \Omega$; and Case 6: $d = 0.1$, $R = 400\ \Omega$.

It is remarked that the experimental results are consistent with the theoretical and simulation results.

References

1. X. Hu, C. Gong, A high gain input-parallel output-series DC-DC converter with dual coupled-inductors. IEEE Trans. Power Electron. **30**(3), 1306–1317 (2015)
2. M.B. Ozkan, P. Karagoz, A novel wind power forecast model: statistical hybrid wind power forecast technique (SHWIP). IEEE Trans. Ind. Inf. **11**(2), 375–387 (2015)
3. W. Song, Y. Zhong, H. Zhang, X. Sun, Q. Zhang, W. Wang, A study of z-source dual-bridge matrix converter immune to abnormal input voltage disturbance and with high voltage transfer ratio. IEEE Trans. Ind. Inf. **9**(2), 828–838 (2013)
4. J. Rodriguez, M.P. Kazmierkowski, J.R. Espinoza, P. Zanchetta, H. Abu-Rub, H.A. Young, C.A. Rojas, State of the art of finite control set model predictive control in power electronics. IEEE Trans. Ind. Inf. **9**(2), 1003–1016 (2013)
5. P. Siano, C. Cecati, H. Yu, J. Kolbusz, Real time operation of smart grids via FCN networks and optimal power flow. IEEE Trans. Ind. Inf. **8**(4), 944–952 (2012)
6. L. Valverde, F. Rosa, C. Bordons, Design, planning and management of a hydrogen-based microgrid. IEEE Trans. Ind. Inf. **9**(3), 1398–1404 (2013)
7. H.L. Ginn, G. Chen, Digital control method for grid-connected converters supplied with non-ideal voltage. IEEE Trans. Ind. Inf. **10**(1), 127–136 (2014)
8. X. Long, J. He, J. Zhou, L. Fang, X. Zhou, F. Ren, T. Xu, A review on light-emitting diode based automotive headlamps. Renew. Sustain. Energy Rev. **41**, 29–41 (2015)
9. M. Rizo, M. Liserre, E. Bueno, F.J. Rodriguez, C. Giron, Voltage control architectures for the universal operation of DPGS. IEEE Trans. Ind. Inf. **11**(2), 313–321 (2015)
10. T.B. Lazzarin, I. Barbi, DSP-based control for parallelism of three-phase voltage source inverter. IEEE Trans. Ind. Inf. **9**(2), 749–759 (2013)
11. M. Bertoluzzo, G. Buja, Development of electric propulsion systems for light electric vehicles. IEEE Trans. Ind. Inf. **7**(3), 428–435 (2011)
12. N. Lu, H. Lin, J. Lu, G. Zhang, A customer churn prediction model in telecom industry using boosting. IEEE Trans. Ind. Inf. **10**(2), 1659–1665 (2014)
13. Q. Yang, J.A. Barria, T.C. Green, Communication infrastructures for distributed control of power distribution networks. IEEE Trans. Ind. Inf. **7**(2), 316–327 (2011)
14. G. Zhang, Z. Li, B. Zhang, D. Qiu, W. Xiao, W.A. Halang, A Z-source half-bridge converter. IEEE Trans. Ind. Electron. **61**(3), 1269–1279 (2014)
15. C. Buccella, C. Cecati, H. Latafat, Digital control of power converters–a survey. IEEE Trans. Ind. Inf. **8**(3), 437–447 (2012)
16. W. Li, Y. Zhao, J. Wu, X. He, Interleaved high step-up converter with winding-cross-coupled inductors and voltage multiplier cells. IEEE Trans. Power Electron. **27**(1), 133–143 (2012)

17. W. Li, L. Fan, Y. Zhao, X. He, D. Xu, High step-up and high efficiency fuel cell power generation system with active clamp flyback-forward converter. IEEE Trans. Ind. Electron. **59**(1), 599–610 (2012)
18. M.A. Al-Saffar, E.H. Ismail, A.J. Sabzali, Family of ZC-ZVS converters with wide voltage range for renewable energy systems. Renew. Energy **56**, 32–43 (2013)
19. J. Duarte, L.R. Lima, L. Oliveira, M. Mezaroba, L. Michels, C. Rech, Modeling and digital control of a single-stage step-up/down isolated PFC rectifier. IEEE Trans. Ind. Inf. **9**(2), 1017–1028 (2013)
20. M. Fu, C. Ma, X. Zhu, A cascaded boost-buck converter for high-efficiency wireless power transfer systems. IEEE Trans. Ind. Inf. **10**(3), 1972–1980 (2014)
21. C. Mario, C. Alfio, A. Rosario, G. Francesco, Soft-switching converter with HF transformer for grid-connected photovoltaic systems. IEEE Trans. Ind. Electron. **57**(5), 1678–1686 (2010)
22. W. Li, Y. He, X. He, Y. Sun, F. Wang, L. Ma, Series asymmetrical half-bridge converters with voltage autobalance for high input-voltage applications. IEEE Trans. Power Electron. **28**(8), 3665–3674 (2013)
23. C.-M. Young, M.-H.Chen, T.-A. Chang, C.-C. Ko, K.-K. Jen, Cascade cockcroft-walton voltage multiplier applied to transformerless high step-up DC-DC converter. IEEE Trans. Ind. Electron. **60**(2), 523–537 (2013)
24. D.S. Wijeratne, G. Moschopoulos, Quadratic power conversion for power electronics: principles and circuits. IEEE Trans. Circuits Syst. I **59**(2), 1967–1979 (2011)
25. G.A.L. Henn, R.N.A.L. Silva, P.P. Praca, L.H.S.C. Barreto, D.S. Oliveira.Jr, Interleaved-boost converter with high voltage gain. IEEE Trans. Power Electron. **25**(11), 2753–2761 (2010)
26. D. Li, P.C. Loh, M. Zhu, G. Feng, F. Blaabjerg, Generalized multicell switched-inductor and switched-capacitor Z-source inverters. IEEE Trans. Power Electron. **28**(2), 837–848 (2013)
27. K.I. Hwu, C.F. Chuang, W.C. Tu, High voltage-boosting converters based on bootstrap capacitors and boost inductors. IEEE Trans. Ind. Electron. **60**(6), 2178–2193 (2013)
28. S. Vighetti, J.P. Ferrieux, Y. Lembeye, Optimization and design of a cascaded DC-DC converter devoted to grid-connected photovoltaic systems. IEEE Trans. Power Electron. **27**(4), 2018–2027 (2012)
29. P.D. Antoszczuk, R. Garcia Retegui, M. Funes, D. Carrica, Optimized implementation of a current control algorithm for multiphase interleaved power converters. IEEE Trans. Ind. Inf. **10**(4), 2224–2232 (2014)
30. C.-T. Pan, C.-F. Chuang, C.-C. Chu, A novel transformerless interleaved high step-down conversion ratio DC-DC converter with low switch voltage stress. IEEE Trans. Power Electron. **61**(10), 5290–5299 (2014)
31. G. Zhang, B. Zhang, Z. Li, A constructing method for power electronics converters based on the graph theory, in *China Power Supply Society 20th Annual Academic Annual Meeting* (2014)
32. G. Zhang, B. Zhang, Z. Li, D. Qiu, L. Yang, W. Halang, A 3-Z-network boost converter. IEEE Trans. Ind. Electron. **62**(1), 278–288 (2015)
33. G. Zhang, H.H.C. Iu, B. Zhang, Z. Li, T. Fernando, S.-Z. Chen, Y. Zhang, An impedance network boost converter with high-voltage gain high voltage-gain. IEEE Trans. Power Electron. **32**(9), 6661–6665 (2017)

Chapter 7
Z-source Half-Bridge Converter

A Z-source half-bridge converter is to be designed and analysed in this chapter. In detail, a Z-network is inserted into the traditional half-bridge converter to increase the input impedance, and balance the mid-voltage of the input capacitors, so that the stringent requirements of the electrochemistry power supply can be satisfied.

7.1 Introduction

Electroplating is a kind of electrochemistry process, whose fundamental operation principle is shown in Fig. 7.1 and described in the following.

The electroplating process is a redox reaction, with fundamental components: two electrodes (+ and -), a DC source (V_d) and the solution, as shown in Fig. 7.1. The purpose of the electroplating is to make the metal ions cover on the surface of the negative electrode evenly and smoothly. However, due to the nonuniform of the solution, the DC voltage direction and the current density should be adjusted from time to time, which requires complicated designs according to different products and processes [1, 2]. With the rapid growth of the demand on electroplating products with very different voltages and duties, there are more stringent requirements on the electrochemical power supplies to provide a broad range of outputs, asymmetrical positive and negative voltages, step waves, recurrent pulses, square waves, triangular waves, and saw-tooth waves [3].

In order to realize the aforementioned functions, the output voltages of the electrochemical power supply are requested to be variable, including variable positive or negative output voltages, and the variable time ratio between positive and negative voltages. For example, in order to realize the smooth electroplating products, the current densities and directions should be varied according to the requests of electroplating technology [1], [2]. Traditionally, the engineer had to compose several cascaded sub-circuits and use complex control methods to generate an overlapped

© Springer International Publishing AG 2018

G. Zhang et al., *Designing Impedance Networks Converters*, Studies in Systems, Decision and Control 119, DOI 10.1007/978-3-319-63655-9_7

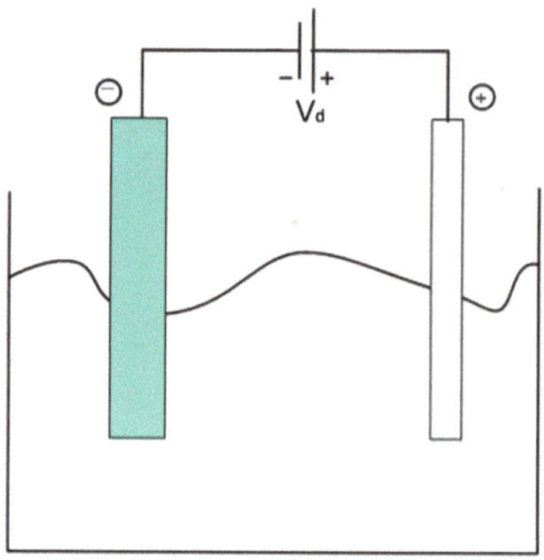

Fig. 7.1 Operation principle of electroplating

waveform of multi output voltages [3, 4]. However, the disadvantages lie in that it is hard to control and regulate the output voltages, and the use of cascaded sub-circuits not only increases the cost and size, but also leads to a more complex, bulky structure and instability of the system. Moreover, the engineers should cascade the corresponding converters with their experience for different electroplating products. Therefore, the traditional design method is on case by case basis and dependant on experts' experience, which cannot fulfil the requirements of modern industrial applications.

In terms of the impedance network matching and the proposed designing methodology, this chapter proposes a novel Z-source half-bridge converter with impedance networks, aiming to the applications in electrochemical systems.

7.2 System Design

DC-AC inverters with abundant output voltages are required in electrochemical systems. However, the traditional DC-AC inverters are limited by the sole output voltage. Thereby, the impedance matching and the systematic design methodology are utilized to realize output impedance matching.

7.2.1 Disadvantages of Traditional Half-Bridge Inverter

A DC-AC inverter with abundant output voltages is required in electrochemical systems, therefore, a half-bridge DC-AC inverter is chosen as the main topology in electrochemical systems considering the economic cost, which is shown in Fig. 7.2. Therein, assume $C_{d1} = C_{d2} = C$.

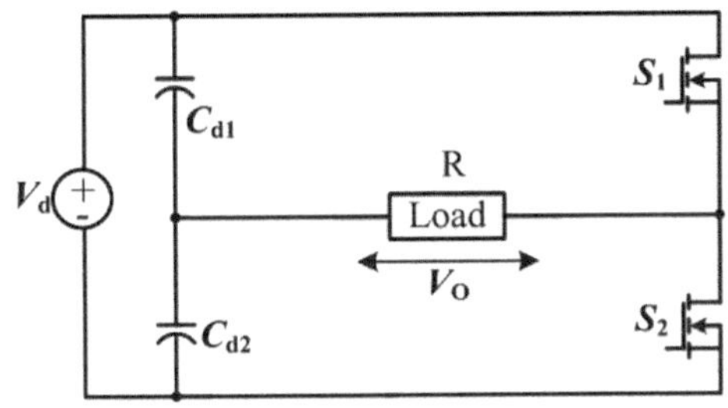

Fig. 7.2 A conventional half-bridge converter

In terms of the input and output impedances analysis methodology, the short-circuited input impedance and average output impedance of the traditional half-bridge converter are expressed by

$$\begin{cases} Z_i(j\omega) = -j\dfrac{1}{\omega C}, \\ Z_o(j\omega) = -j\dfrac{1}{2\omega C}. \end{cases} \tag{7.1}$$

It is noted from (7.1) that the imaginary parts of the input and output impedances are both negative. Therefore, similar to voltage source inverters, it has some disadvantages, i.e. the shoot-through problems, low output voltage, thus cannot fulfill the requirements of electrochemical systems, and the inapplicability to capacitive loads.

7.2.2 Impedance Matching of Traditional Half-Bridge Inverters

In order to overcome the disadvantages listed above, the X-shaped two-port impedance network type in Table 5.2 is chosen as the impedance network. Then, embedding the Z-network between the capacitors and the inverter bridge in Fig. 7.2 results in the impedance-network half-bridge converter as shown in Fig. 7.3.

Therein, an LC impedance network, consisting of capacitors C_1 and C_2 and inductors L_1 and L_2, is integrated into a traditional half-bridge converter, consisting of capacitors C_{d1} and C_{d2}, switches S_1 and S_2, and diode D, which is used to prevent the current from flowing back to the source. Moreover, the use of the inductors in Z-network is to avoid strong current in the circuit when the switches are in the shoot-through state [5].

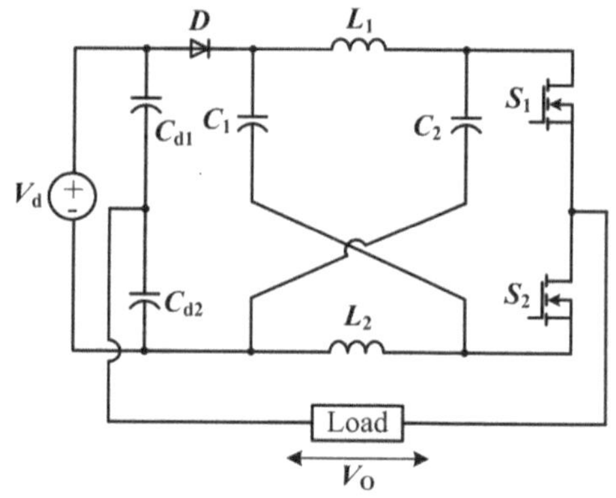

Fig. 7.3 Z-source half-bridge converter

7.2.3 *Calculation of Input and Output Impedances*

One can obtain the transmission parameters matrix of X-shaped two-port impedance network in terms of the equations in Table 5.2, substituting the corresponding parameters into the input output impedances equations in Tables 5.4 and 5.5. Then, it is easy to prove that they all satisfy the impedance matching conditions, i.e. the short-circuited input impedance is unequal to 0 to realize input impedance matching, while the output impedance can be adapted to match the load impedance for realizing the desired abundant output voltage. The process is omitted here because it is similar as the one of typical Z-source converter.

7.3 Operational Status Analysis

For simplicity, it is assumed that (a) all the components are ideal; (b) the dead time in the driven pulses is ignored; (c) $L_1 = L_2$ and $C_1 = C_2$ in the Z-network; (d) C_1, C_2, C_{d1} and C_{d2} are large enough; and (e) the free-wheeling diodes of the switches are ignored in the analysis since the load characteristic of the electrochemical solution is resistance or resistance with a small capacitance.

Denote the duties of the switches S_1 and S_2 by d_1 and d_2, respectively. The proposed converter performs differently in two cases, i.e. Case 1: $d_1 + d_2 \leq 1$ and Case 2: $d_1 + d_2 > 1$.

(1) Case 1

In this case, S_1 and S_2 are not on at the same time, which is named as the non-shoot-through state.

There are three modes corresponding to the states of the switches.

1. Fig. 7.4a shows an equivalent circuit of the first mode, for which the S_1 is on and S_2 is off, where the current flows out of the source, through the diode,

Fig. 7.4 Equivalent circuits in Case 1

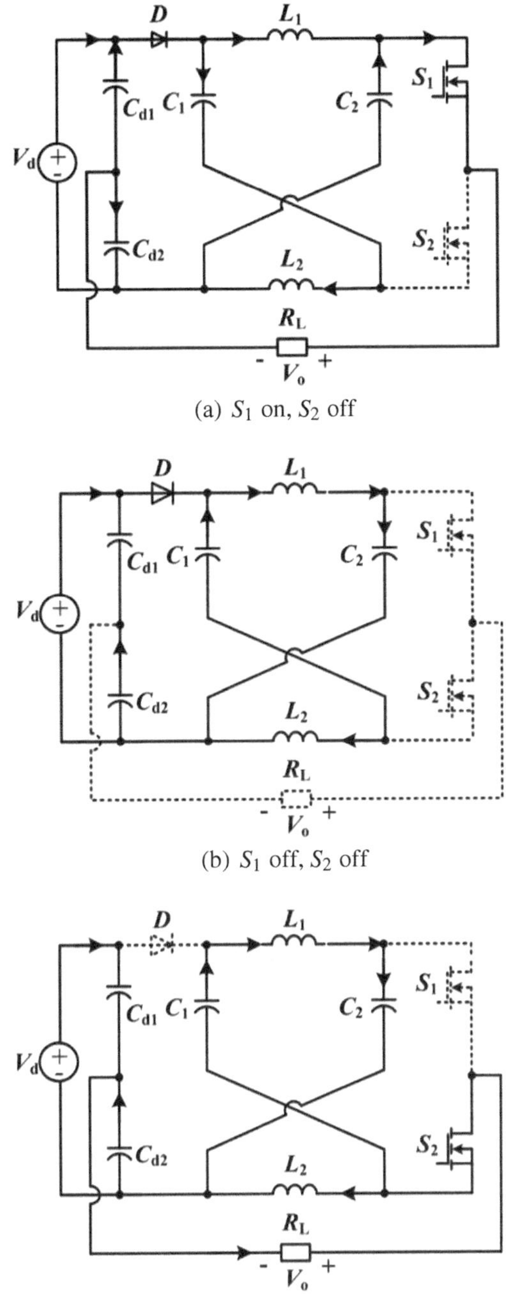

(c) S_1 off, S_2 on

the Z-network, S_1, and then back to the source. The arrows indicate the current direction.

2. Fig. 7.4b illustrates an equivalent circuit of the second mode, for which S_1 and S_2 are off, where the current also flows out of the source, through the diode, the Z-network, and back to the source. It is noted that here is no output.
3. Fig. 7.4c depicts an equivalent circuit of the third mode, for which S_2 is on and S_1 is off, where the diode undertakes a negative voltage, and thus, turns off. The current flows out of the source, through the load, S_2, Z-network, and then back to the source. Besides, the current direction is also indicated. The operation process for this case is similar to the traditional one for half-bridge converters, which is not detailed here.

(2) Case 2

In this case, the behavior of the switches in the circuit leads to three modes within a switching period T, which correspond to three linear equivalent circuits: Mode 1, when S_1 and S_2 are on; Mode 2, when S_1 is on and S_2 is off; and Mode 3, when S_1 is off and S_2 is on, as shown in Fig. 7.5a, b and c, respectively.

Denote t_0 the initial time of one period, t_1 the mode transition instant from Mode 1 to Mode 2, i.e. $t_1 = t_0 + (d_2 + d_1 - 1)T$, t_2 the mode transition instant from Mode 2 to Mode 3, i.e. $t_2 = t_1 + (1 - d_2)T$, and $t_3 = T$ the end of the period.

In the steady state of the converter, its operational process in a switching period is analysed in the following and the output voltage v_o will be deduced in each mode.

Mode 1: $t \in [t_0, t_1]$

As shown in Fig. 7.5a, in loops 1 and 2, capacitors C_1 and C_2 discharge the energy to inductors L_1 and L_2, thereafter, i_{L_1} and i_{L_2} increase. Thus, L_1 and L_2 store the energy, and one has

$$\begin{cases} v_{L_1} = v_{C_1} , \\ v_{L_2} = v_{C_2} , \end{cases} \tag{7.2}$$

where i_{L_1}, i_{L_2}, v_{L_1}, v_{L_2}, v_{C_1} and v_{C_2} are the currents of L_1, L_2, and the voltages of L_1, L_2, C_1 and C_2, respectively.

The voltage of diode D is $-(v_{C_1} + v_{C_2} - V_d)$, so D undertakes negative voltage stress, and thus, turns off. The energy of C_2 is delivered to the load R_L and C_{d2} through the C_2-R_L-C_{d2} loop, so C_{d2} charges and C_{d1} discharges.

In terms of the C_2-R_L-C_{d2} loop, the output voltage of the converter reads

$$v_o = v_{C_2} - v_{C_{d2}} , \tag{7.3}$$

where $v_{C_{d2}}$ is the voltage of C_{d2}.

Mode 2: $t \in [t_1, t_2]$

As shown in Fig. 7.5b, S_1 is on and S_2 is off. In loop 1, the source V_d and L_1 discharge the energy to C_2, so that v_{C_2} increases. In loop 2, the source V_d and L_2 discharge the energy to C_1, thereafter, v_{C_1} increases. Then, the energy of C_2 is delivered to the load R_L and C_{d2} through the C_2-R_L-C_{d2} loop, so C_{d2} charges and C_{d1} discharges. From loop 1, one has

Fig. 7.5 Equivalent circuits in Case 2

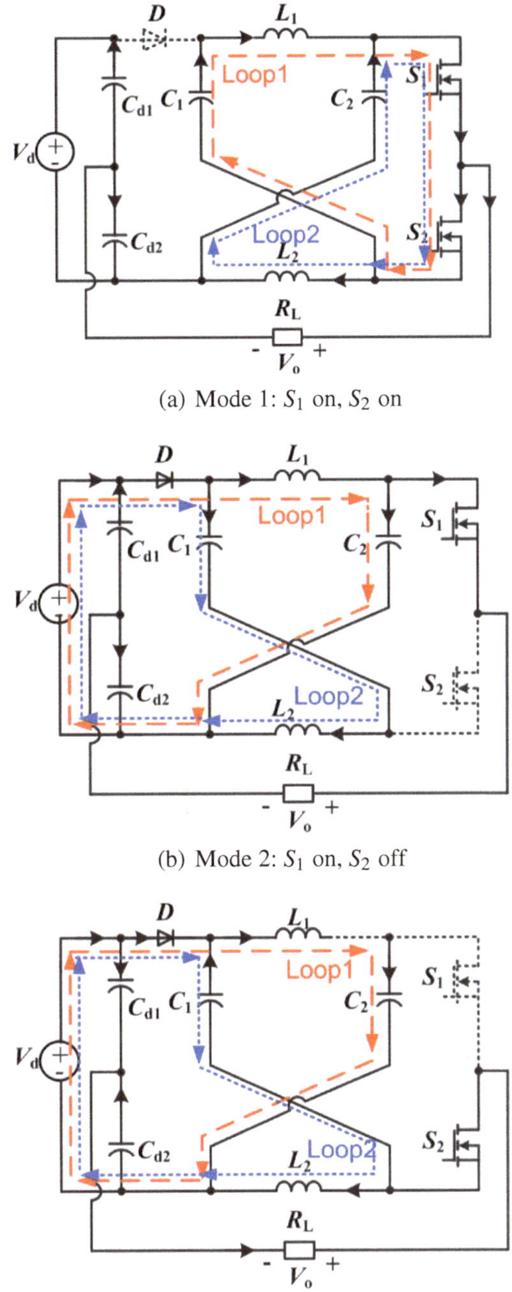

(a) Mode 1: S_1 on, S_2 on

(b) Mode 2: S_1 on, S_2 off

(c) Mode 3: S_1 off, S_2 on

$$v_{L_1} = V_d - v_{C_2}\ . \tag{7.4}$$

In terms of the C_2-R_L-C_{d2} loop, the output voltage of the converter is same as (7.3).

Mode 3: $t \in [t_2, t_3]$

In Fig. 7.5c, S_1 is off and S_2 is on. In loop 1, the source V_d and L_1 discharge the energy to C_2, thus, v_{C_2} increases. Similarly, in loop 2, V_d and L_2 discharge the energy to C_1, thus, v_{C_1} increases. The energy of L_2 and C_{d2} is delivered to R_L through the L_2-C_{d2}-R_L loop, so C_{d2} discharges and C_{d1} charges. In terms of loop 1, one has the same Eq. (7.4).

In terms of the V_d-D-C_1-R_L-C_{d2} loop, the output of the voltage is

$$v_o = -(v_{C_{d2}} + v_{C_1} - V_d)\ . \tag{7.5}$$

As a result, v_o can be deduced as follow.

The voltage-second balance characteristic of L_1 leads to

$$\int_0^T v_{L_1} dt = 0\ . \tag{7.6}$$

Substituting (7.2) and (7.4) into (7.6) leads to

$$(d_2 + d_1 - 1)T V_{C_1} + (2 - d_2 - d_1)T(V_d - V_{C_2}) = 0\ . \tag{7.7}$$

Assume that $L_1 = L_2$, $C_1 = C_2$, and C_1 and C_2 are large enough. Due to the structural symmetry of the Z-network, (7.7) can be rewritten as

$$V_{C_1} = V_{C_2} = \frac{2 - d_1 - d_2}{3 - 2(d_1 + d_2)} V_d\ . \tag{7.8}$$

The ampere-second property of C_{d2} implies

$$\int_0^T i_{C_{d2}} dt = 0\ , \tag{7.9}$$

where $i_{C_{d2}}$ is the current of C_{d2}.

Denote the voltage and current of C_{d1} by $v_{C_{d1}}$ and $i_{C_{d1}}$, respectively. It is known from Fig. 7.5 that $v_{C_{d1}} + v_{C_{d2}} = V_d$. Denote the errors of $v_{C_{d1}}$ and $v_{C_{d2}}$ by $\Delta v_{C_{d1}}$ and $\Delta v_{C_{d2}}$, respectively, and due to V_d being a constant, one has $\Delta v_{C_{d1}} = -\Delta v_{C_{d2}}$, and straightforwardly, $i_{C_{d1}} = i_{C_{d2}}$ in terms of $i = C\frac{du}{dt}$.

Moreover, from $i_o = i_{C_{d1}} + i_{C_{d2}}$, one has $i_{C_{d2}} = \frac{i_o}{2}$, where i_o is the current of the load; thereafter, (7.9) can be rewritten as

$$\frac{(V_{C_2} - V_{C_{d2}})}{2R_L} d_1 T + \frac{-(V_{C_2} + V_{C_{d2}} - V_d)}{2R_L}(1 - d_1)T = 0\ , \tag{7.10}$$

and it follows with that

$$V_{C_{d2}} = (2V_{C_2} - V_{\mathrm{d}})d_1 - V_{C_2} + V_{\mathrm{d}} \,. \tag{7.11}$$

When switch S_1 is on, substituting (7.8) and (7.11) into (7.3) results in the positive output of the converter V_{p} as

$$V_{\mathrm{p}} = v_{\mathrm{o}} = V_{C_2} - v_{C_{d2}} = \frac{(1-d_1)}{3-2(d_1+d_2)}V_{\mathrm{d}} \,. \tag{7.12}$$

When the switch S_2 is on and S_1 is off, substituting (7.8) and (7.11) into (7.5) leads to the negative output of the converter V_{n} as

$$V_{\mathrm{n}} = v_{\mathrm{o}} = V_{\mathrm{d}} - V_{C_2} - V_{C_{d2}} = -\frac{d_1}{3-2(d_1+d_2)}V_{\mathrm{d}} \,. \tag{7.13}$$

According to (7.12) and (7.13), the relationships of d_1, d_2 and $V_{\mathrm{o}}/V_{\mathrm{d}}$ are drawn in Fig. 7.6. Therein, $V_{\mathrm{o}}/V_{\mathrm{d}}$ increases dramatically as d_1+d_2 is about 1.5, which is zoomed-in in Fig. 7.7. It is remarked from Figs. 7.6 and 7.7 that the novel converter can generate abundant output voltages by adjusting d_1 and d_2.

When $V_{\mathrm{o}}/V_{\mathrm{d}} < 1$, the converter functions as a Buck converter; otherwise, the converter acts as a boost converter; therefore, it is a buck-boost converter. By controlling the duty of the switches, specified output voltages can be obtained, including the buck-boost voltages, asymmetric and symmetric, positive and negative peak output voltages, and the time ratio between positive and negative voltages.

Additionally, according to (7.12) and (7.13), the values of V_{p} and V_{n} are not equal when $d_1 \neq 0.5$, but they are equal when $d_1 = 0.5$, which will be explained as below.

Firstly, when $d_1 = 0.5$, the key waveforms of the Z-source half-bridge converter in Case 2 is drawn in Fig. 7.8 according to the analysis for three modes, where Q_{S_1} and Q_{S_2} stand for the driving voltages of switches (S_1 and S_2), i_d the current of diode (D), i_{L_1} and i_{L_2} the currents of inductors (L_1 and L_2), v_{C_1}, v_{C_2}, $v_{C_{d1}}$ and $v_{C_{d2}}$ the voltages of the capacitors (C_1, C_2, C_{d1} and C_{d2}), and v_{o} the output voltage, respectively. Additionally, Modes 1, 2, and 3 are distinguished in red, blue, and green colors, respectively. The limited output voltages of the traditional half-bridge converter, $V_{\mathrm{d}}/2$ and $-V_{\mathrm{d}}/2$, are marked at the output voltage waveform v_{o}, and it is shown that the output voltages of the proposed converter can exceed the limited one.

Secondly, the corresponding waveforms for $d_1 \neq 0.5$ are shown in Fig. 7.9. It is remarked that the output voltage v_{o} in Fig. 7.9 is different from that in Fig. 7.8; and the positive and negative values of v_{o} in Fig. 7.8 are symmetrical, but they are asymmetrical in Fig. 7.9. This means that the proposed converter generates many kinds of output voltages, fulfilling the requirements of the electrochemical power supply, such as, various positive or negative output voltages, and the regulated duration at negative or positive output voltage, which has prominent advantages over traditional methods by using complicated control methods and multiple cascaded sub-circuits.

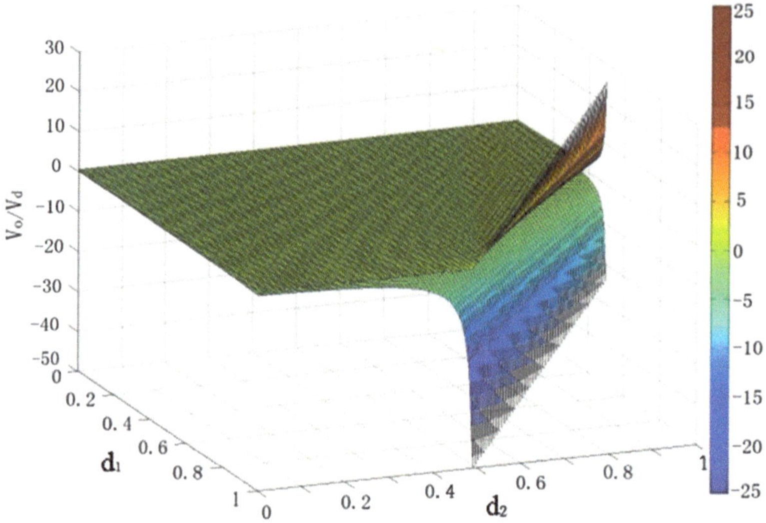

Fig. 7.6 Relationship figure of d_1, d_2, and V_o/V_d

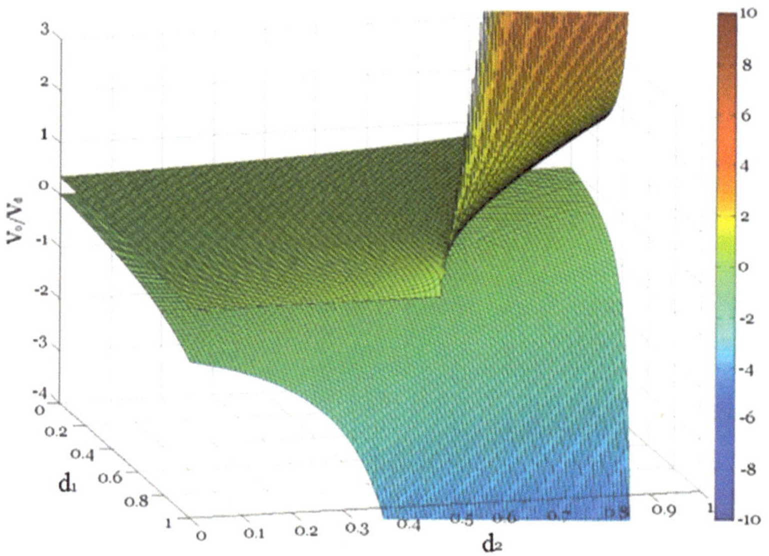

Fig. 7.7 Zooming in of Fig. 7.6

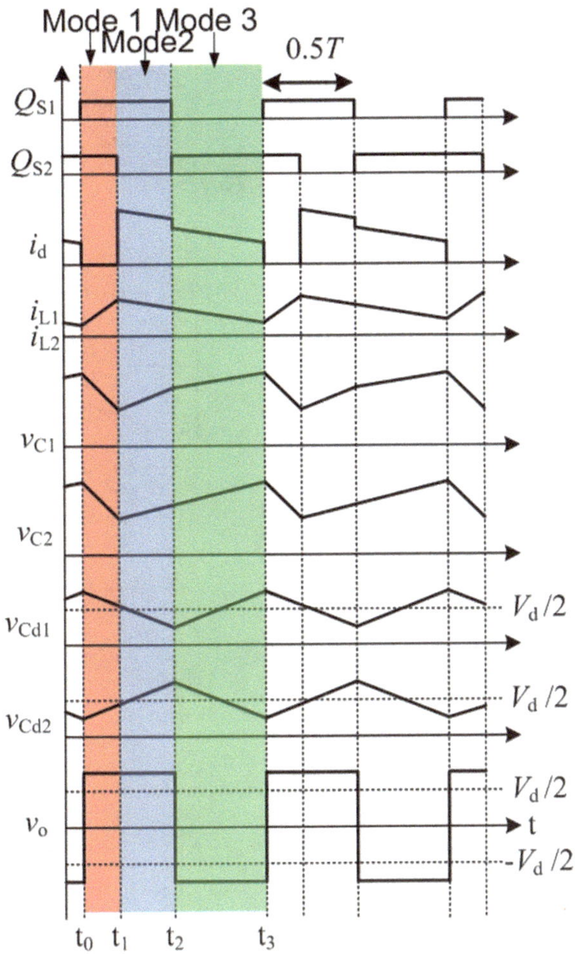

Fig. 7.8 Key waveforms of the Z-source Half-bridge converter in Case 2 when $d_1 = 0.5$ and $d_2 = 0.7$

Moreover, the efficiency of the converter η is given by

$$\begin{aligned}\eta &= \frac{P_{\text{out}}}{P_{\text{in}}}\\ &= \frac{d_1\dfrac{v_{\text{p}}^2}{R} + (1-d_1)\dfrac{v_{\text{n}}^2}{R}}{V_{\text{d}}I_{\text{av}}}\\ &= \frac{d_1 v_{\text{p}}^2 + (1-d_1)v_{\text{n}}^2}{RV_{\text{d}}I_{\text{av}}}\\ &= \frac{d_1(1-d_1)V_{\text{d}}}{(3-2(d_1+d_2))^2 RI_{\text{av}}},\end{aligned} \tag{7.14}$$

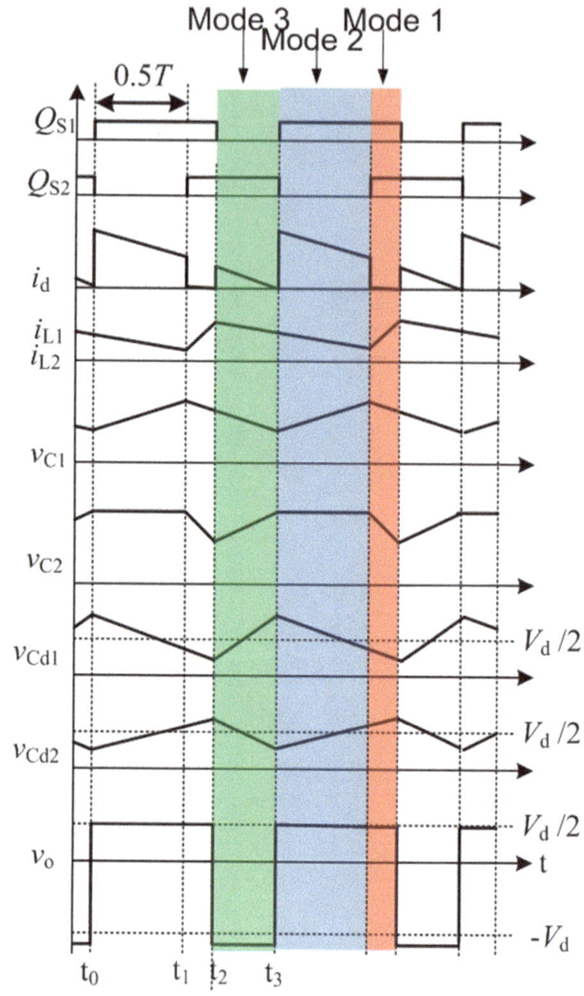

Fig. 7.9 Waveforms of the Z-source half-bridge converter in Case 2 when $d_1 = 0.7$ and $d_2 = 0.5$

where $P_{out} = \dfrac{d_1 v_p^2 + (1 - d_1)v_n^2}{R}$, $P_{in} = V_d I_{av}$, I_{av} are the output power, the input power, and the average input current, respectively. Here, the conduction and switching loss is taken into account, and indicated in $P_{in} - P_{out}$.

7.4 Mid-Point Balance of Input Capacitors

The stability of the mid-point voltage in the converter plays a key rôle for the system's stability. The mid-point voltages of the input capacitors in the conventional converter and the proposed converter will be analysed and compared in this section.

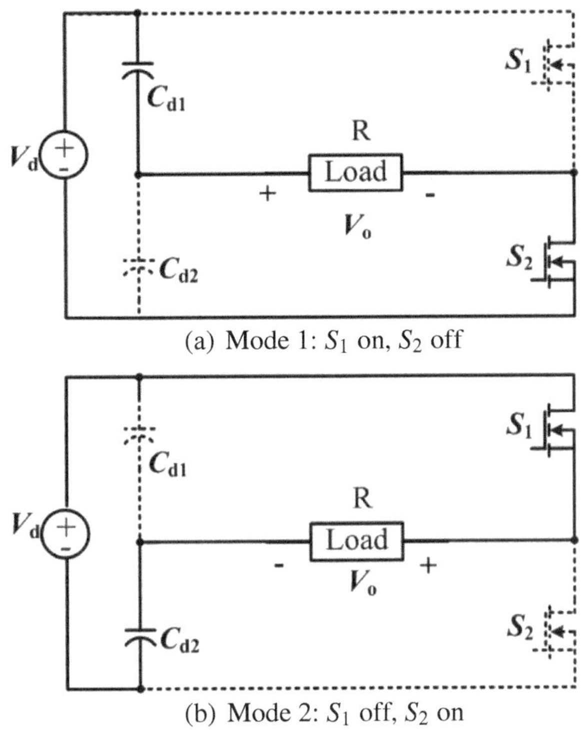

Fig. 7.10 Equivalent circuits of the conventional half-bridge converter

7.4.1 Mid-Point Voltage in Conventional Half-Bridge Converters

In conventional half-bridge converters, there are always some problems caused by the mid-point unbalance of the input capacitor voltage. In this subsection, the mid-point voltage in conventional half-bridge converters will be analysed and the fluctuation equation of the mid-point voltage will be derived.

Figure 7.10 shows the equivalent circuits of that in Fig. 7.2.

In a switching period, S_1 is on and S_2 is off as $t \in [0, d_1 T]$, while S_1 is off and S_2 is on as $t \in [d_1 T, T]$.

Denote the initial voltage of C_{d2} by $V_{C_{d2}0}$. In terms of the Kirhhoff's Voltage Law (KVL), $v_{C_{d2}}$ can be derived in frequency domain as

$$v_{C_{d2}}(s) = \frac{V_d}{s} - \frac{V_d - V_{C_{d2}0}}{s + \dfrac{1}{C_{d2}R}}. \tag{7.15}$$

Employing Laplace Inverse Transformation to (7.15) results in

$$v_{C_{d2}}(t) = V_{\mathrm{d}} - (V_{\mathrm{d}} - V_{C_{d2}0})e^{-\frac{t}{C_{d2}R}} . \tag{7.16}$$

Denote the maximal and the minimal voltages of $v_{C_{d2}}$ by $v_{C_{d2}\max}$ and $v_{C_{d2}\min}$, respectively, and the maximal fluctuation of $v_{C_{d2}}$ by ΔV. Following (7.16), one has

$$\Delta V = v_{C_{d2}\max} - v_{C_{d2}\min} \tag{7.17}$$

$$= (V_{\mathrm{d}} - V_{C_{d2}0})(1 - e^{-\frac{d_1 T}{C_{d2}R}}) . \tag{7.18}$$

In the high-frequency power supply, T is normally very small, and the input capacitance C_{d2} and load are quite large in electrochemical application, which implies that $d_1 T$ is much smaller than $C_{d2}R$, then (7.17) can be approximated by

$$\Delta V \approx \frac{d_1 t}{C_{d2}R}(V_{\mathrm{d}} - V_{C_{d2}0}) . \tag{7.19}$$

7.4.2 *Mid-Point Voltage in Z-source Half-Bridge Converters*

It is described in [6] that the input part can be regarded as a DC voltage source or a DC current source due to the Z-network. Similarly, the output part of the Z-network can be treated as a DC current source. Hence, the equivalent circuits are derived as follows.

The differential equation of the circuit shown in Fig. 7.11a can be described as

$$C_{d2}\frac{dv_{C_{d2}}}{dt} = I_{\mathrm{p}} , \tag{7.20}$$

where I_{p} is the current of the constant current source.

Integrating both sides of (7.20) leads to

$$v_{C_{d2}}(t) = V_{C_{d2}0} + \int \frac{I_{\mathrm{p}}}{C_{d2}}dt . \tag{7.21}$$

Denote the maximal fluctuation of $v_{C_{d2}}$ as shown in Fig. 7.11 by ΔV_{Z}. Then, from (7.21), one has

$$\Delta V_{\mathrm{Z}} = v_{C_{d2}\max} - v_{C_{d2}\min} \tag{7.22}$$

$$= \int_0^{d_1 T} \frac{I_{\mathrm{p}}}{C_{d2}}dt = d_1 T\frac{I_{\mathrm{p}}}{C_{d2}} . \tag{7.23}$$

I_{p} can be derived by Kirhhoff's Current Law (KCL) as

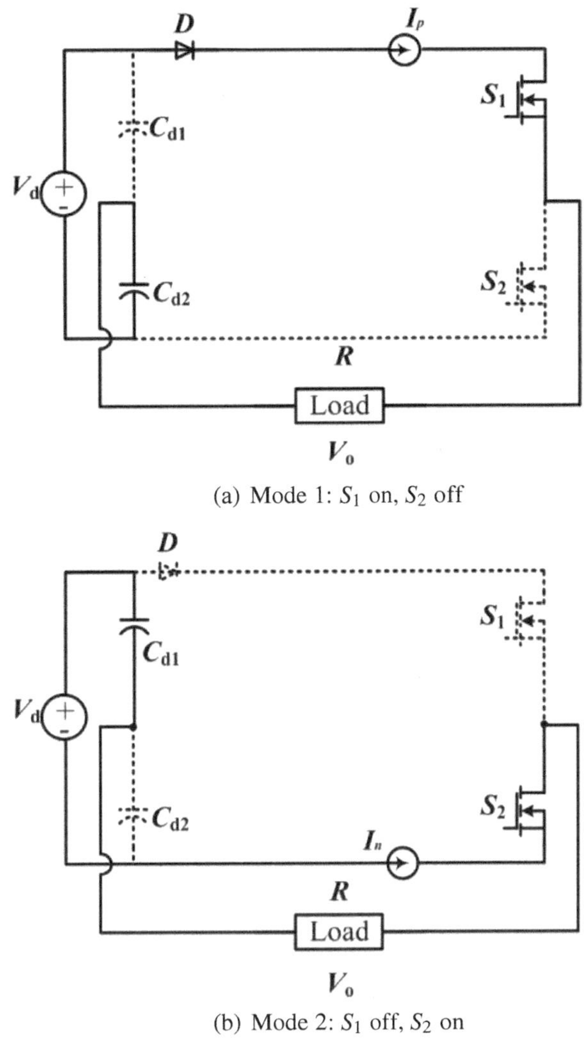

Fig. 7.11 Equivalent circuits of the Z-source half-bridge converter

$$I_p = \frac{V_d - V_{C_{d2}0} - V_{I_p}}{R}, \tag{7.24}$$

where V_{I_p} is the voltage of the constant current source.

Substituting (7.24) into (7.22) leads to

$$\Delta V_Z = d_1 T \frac{V_d - V_{C_{d2}0} - V_{I_p}}{C_{d2} R}. \tag{7.25}$$

Therein, the ratio of ΔV_Z to ΔV can be derived from (7.19) and (7.25) as

$$\frac{\Delta V_Z}{\Delta V} = \frac{V_d - V_{C_{d2}0} - V_{I_p}}{V_d - V_{C_{d2}0}} \times 100\% . \tag{7.26}$$

It is obvious that $\Delta V_Z < \Delta V$, if V_{I_p} is positive; and the smaller $\frac{\Delta V_Z}{\Delta V}$ is, the smaller the ripple in the proposed converter, consequently, the more stable the proposed converter is. $\frac{\Delta V_Z}{\Delta V}$ will become very small, or even zero, if V_{I_p} is very close to the value of $V_d - V_{C_{d2}0}$, and V_{I_p} can be designed by the parameters of the Z-network.

It is remarked that the proposed converter is more stable than the traditional one with regard to the problem of the input capacitor stability.

7.5 Parameters Determination

7.5.1 Parameters of Capacitors

Normally, the design of the capacitor is to determine the rated voltage and capacitance with a permitted fluctuation range, $x_C\%$ (x_C is pre-assigned), a given output voltage V_o, a given output current I_o, and a given switching period T.

From (7.8), (7.12), and (7.13), one has

$$v_{C_2} = \frac{2 - d_1 - d_2}{d_1} V_o, \text{ when } (S_1) = (\text{on}), \tag{7.27}$$

and

$$v_{C_2} = \frac{2 - d_1 - d_2}{1 - d_1} V_o, \text{ when } (S_1, S_2 = (\text{off}, \text{on})). \tag{7.28}$$

In terms of Kirchhoff's current law (short for KCL), the equations of the connected nodes of L_2-C_1-S_2 in Fig. 7.5a, L_1-C_2-S_1 in Fig. 7.5b, and L_2-C_1-S_2 in Fig. 7.5c can be derived as

$$\begin{cases} i_{L_2} = i_{C_1} + i_o, \text{ when } (S_1, S_2) = (\text{on}, \text{on}), \\ i_{L_1} = i_{C_2} + i_o, \text{ when } (S_1, S_2) = (\text{on}, \text{off}), \\ i_{L_2} = i_o - i_{C_1}, \text{ when } (S_1, S_2) = (\text{off}, \text{on}). \end{cases} \tag{7.29}$$

Denote the root-mean-square (rms) currents of L_2 and C_2 by I_{L_2} and I_{C_2}, respectively. Then, from (7.29), one has

$$I_{C_2} \approx I_{L_2} = \frac{I_o}{2}. \tag{7.30}$$

7.5.1.1 Determination of the Rated Voltage

The range of v_{C_2} is determined by (7.27) and (7.28). Thereby, the rated voltage of C_2 can be determined by the maximal V_{C_2M}. Considering the safety margin, the rated voltage of C_2 is normally set between $1.5V_{C_2M}$ and $2V_{C_2M}$.

7.5.1.2 Determination of the Rated Capacitance

The ripples of the capacitors have great influence on the stability of the converter, whose permitted fluctuation range can be used to design the capacitance.

Then, the capacitors in Z-network can be determined according to the differential equation of capacitors

$$C_2 = \frac{i_{C_2}dt}{dv_{C_2}} . \tag{7.31}$$

The high harmonic frequency of the capacitance is nearly equal to the switching frequency of the converter, as shown in Fig. 7.8, namely

$$dt \approx (d_1 + d_2 - 1)T . \tag{7.32}$$

Denote the permitted error of V_{C_2M} by dv_{C_2}, according to the permitted fluctuation range $x_C\%$, dv_{C_2} is expressed as

$$dv_{C_2} = x_C\% V_{C_2M} . \tag{7.33}$$

Substituting (7.30), (7.32) and (7.33) into (7.31) leads to

$$C_2 = \frac{I_o(d_1 + d_2 - 1)T}{2x_C\% V_{C_2M}} . \tag{7.34}$$

Therein, the range of the capacitance can be calculated, and the maximal is taken as the rated capacitance.

7.5.2 *Parameters of Inductors*

Similar to the parameter design of the capacitor, the parameter design of the inductor is to determine the rated current and inductance with a permitted fluctuation range $x_L\%$ (x_L is pre-assigned), a given output voltage V_o, a given output current I_o, and a given switching period T.

7.5.2.1 Determination of the Rated Current

I_{L_2} can be determined by (7.30). Considering the safety margin, the rated current of L_2 is normally taken as $2I_{L_2}$.

7.5.2.2 Determination of the Rated Inductance

The ripples of the inductors also have great influence on the stability of the converter; therefore, the inductance can be designed in terms of the permitted ripples.

The inductors in the Z-network can be determined according to the differential equation of inductances

$$L_2 = \frac{v_{L_2} dt_L}{di_{L_2}}. \tag{7.35}$$

In the L_1-C_2-L_2-C_1 loop, the KVL equation can be expressed as $v_{L_2} + v_{C_1} = v_{L_1} + v_{C_2}$. In the Z-network, the rms voltages of C_1, C_2, L_1, and L_2 are denoted by V_{C_1}, V_{C_2}, V_{L_1} and V_{L_2}, respectively; and one has $v_{C_1} \approx V_{C_2}$ and $V_{L_2} \approx V_{L_1}$. Thereby, the maximal of v_{L_2} is derived as

$$V_{L_2\mathrm{M}} \approx V_{C_2\mathrm{M}}. \tag{7.36}$$

The high harmonic frequency of the inductor is nearly equal to the switching frequency of the converter, as shown in Fig. 7.8, so the time interval dt_L in (7.35) can be obtained as

$$dt_L \approx (d_1 + d_2 - 1)T. \tag{7.37}$$

Denote the permitted error of I_{L_2} by di_{L_2}. According to the permitted fluctuation range $x_L\%$, di_{L_2} is expressed as

$$di_{L_2} = x_L\% I_{L_2}. \tag{7.38}$$

Substituting (7.27), (7.28), (7.30), (7.36), (7.37) and (7.38) into (7.35) leads to the inductance of L_2 as

$$L_2 = \frac{2V_{C_2\mathrm{M}}(d_1 + d_2 - 1)T}{x_L\% I_\mathrm{o}}. \tag{7.39}$$

7.6 Simulations and Experiments

To verify the feasibility and validity of the proposed converter, Simulink software is applied for the simulation of the converter.

The pre-assigned parameters are as follows, $x_C\% = 1\%$, $x_L\% = 10\%$, $V_d = 48$ V, $V_o = 100$ V, $I_o = 10$ A, and $T = 20$ μs. According to the design, the parameters of the converter can be calculated: $C_1 = C_2 = 482.5$ μF, $L_1 = L_2 = 105.5$ μH. However, in practice, the parameters can be chosen as follows: $C_1 = C_2 = 470$ μF, $L_1 = L_2 = 100$ μH.

The simulation results are shown in Figs. 7.12 and 7.13, which are consistent to theoretical analyses shown in Figs. 7.8 and 7.9.

A prototype is built as shown in Fig. 7.14, and the parameters are chosen as follow: $C_{d1} = C_{d2} = 470$ μF, $C_1 = C_2 = 470$ μF, $L_1 = L_2 = 100$ μH, $R = 100$ Ω, and $T = 20$ μs.

The main circuit is in the left side in Fig. 7.14, composed of L_1, L_2, C_1, C_2, switches (Type: IRFP450) and the resistive load R; while the driving circuit is in the

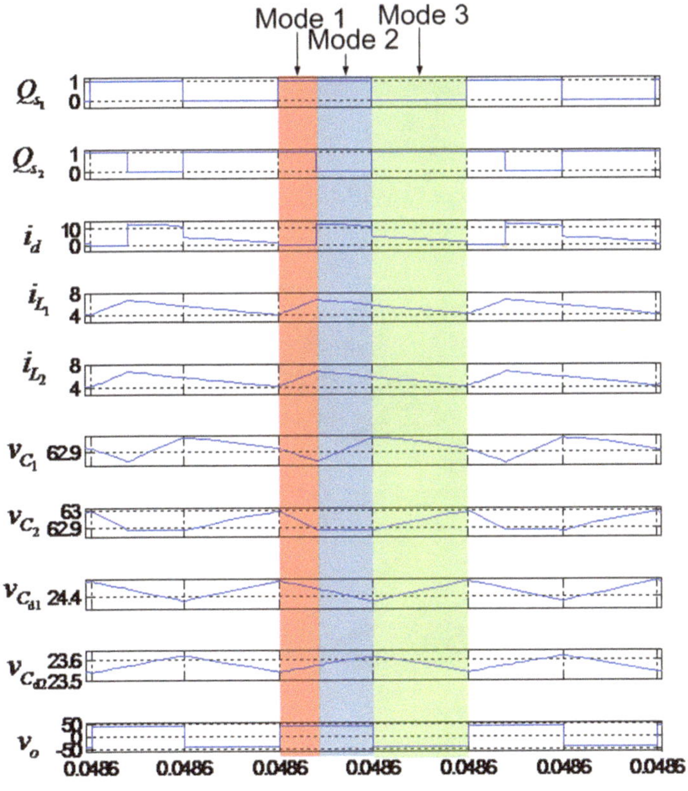

Fig. 7.12 Simulation waveforms when $d_1 = 0.5$ and $d_2 = 0.7$

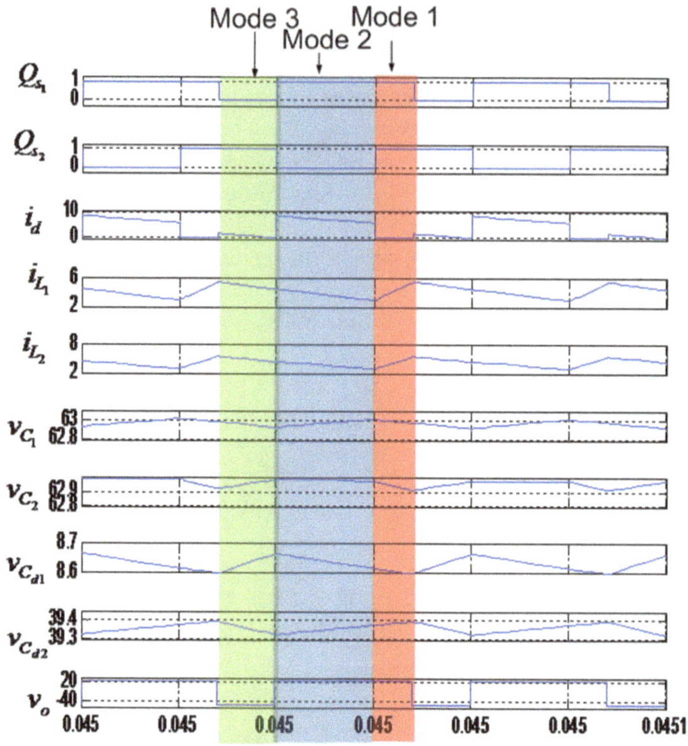

Fig. 7.13 Simulation waveforms when $d_1 = 0.7$ and $d_2 = 0.5$

Fig. 7.14 Prototype of the proposed converter

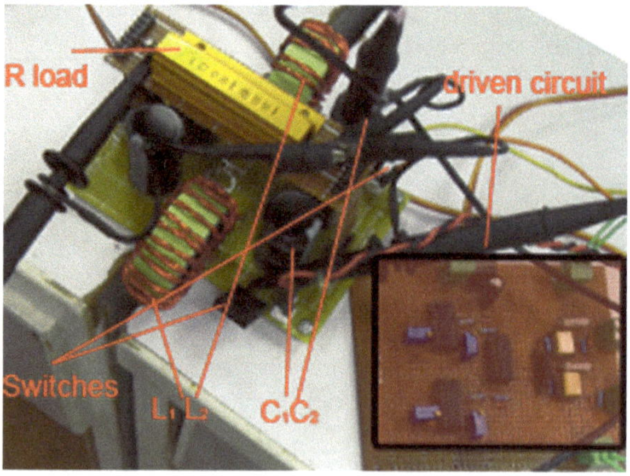

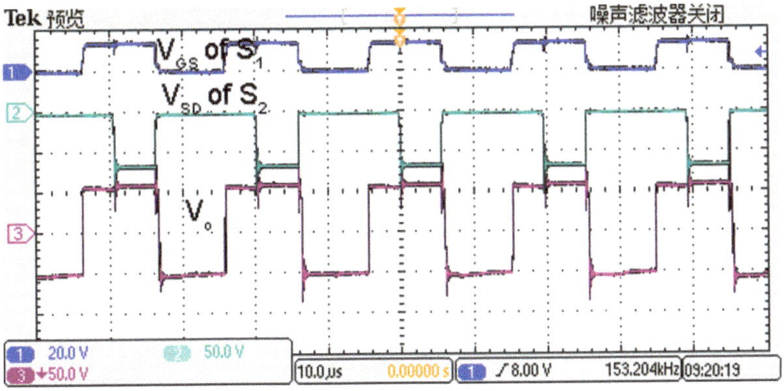

Fig. 7.15 Experimental waveforms in shoot-through case ($d_1 = 0.5, d_2 = 0.7$)

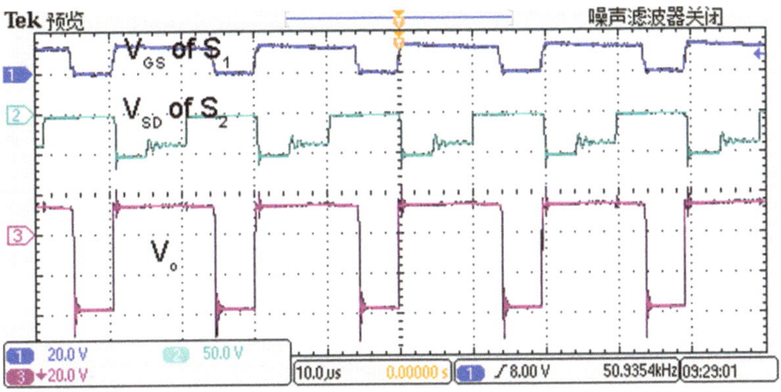

Fig. 7.16 Experimental waveforms in shoot-through case ($d_1 = 0.7, d_2 = 0.5$)

right side, composed of two SG3525 ICs being applied to generated two synchronous overlapped driving signals, and TLP250 ICs being used to drive the switches, whose working frequency and duties can be adapted by the adjustable resistors.

The waveforms of the converter at $d_1 = 0.5$ and $d_2 = 0.7$ with an input voltage 40 V are shown in Fig. 7.15. Therein, the upper waveform refers to V_{GS}(Gate-Drain voltage) of the switch S_1; the middle one is V_{SD}(Source-Drain voltage) of the switch S_2, which is not, but can be synchronized to the driving waveform of S_2; the lower one is the output voltage of the load R, whose negative and positive output voltages are symmetric, and they are all about 50 V. This verifies the analytical and simulation results.

Figure 7.16 depicts the experimental waveforms of the converter when $d_1 = 0.7$ and $d_2 = 0.5$. Therein, the negative and positive output voltages are asymmetric, the positive one is about 20 V, which is nearly equal to $V_d/2$, and the width is $d_1 T$; while the negative one is 40 V, which is much larger than $V_d/2$. The experimental results are also consistent with the simulation results.

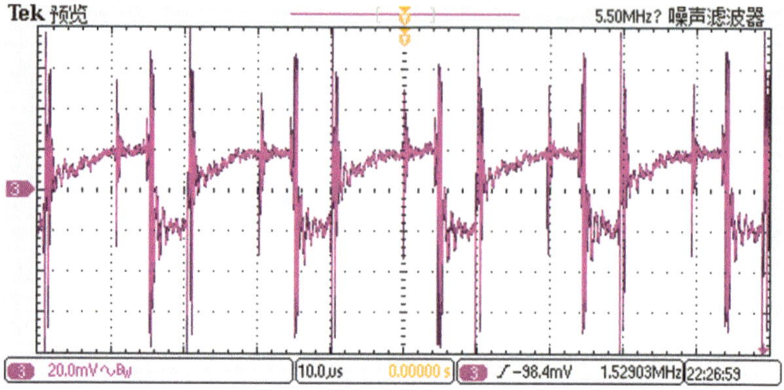

Fig. 7.17 The ripple experiment waveform of $v_{C_{d2}}$ in the proposed converter

Fig. 7.18 Comparison between the experimental and the estimation efficiency

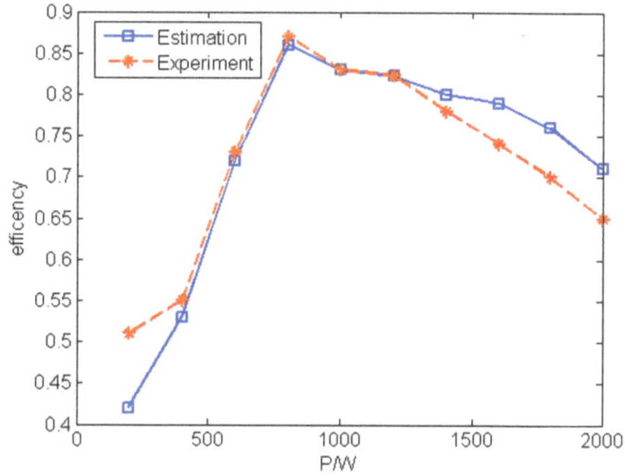

In order to verify that the proposed converter with a balanced mid-point voltage, the experimental result is shown in Fig. 7.17. Therein, the ripples of $v_{C_{d2}}$ in the proposed converter have maximal peak-to-peak values just about 98.4 mV.

Substituting $P_{\text{out}} = 200 \cdot N$ W($N = 1, 2, \ldots 10$) into (7.14), one can obtain the corresponding efficiencies in theory, which are compared with the corresponding experimental efficiencies in the Fig. 7.18. It is found that the experimental results are nearly consistent with the theoretical results, moreover, the efficiency of the proposed converter reaches the high efficiency when the power is 80% rated power.

References

1. X. Hu, Z.Y. Ling, X.H. He, S.S. Chen, Controlling transmission spectra of photonic crystals under electrochemical oxidization of aluminum. J. Electrochem. Soc. **156**(5), C176–C179 (2009)
2. X. Hu, Z.Y. Ling, T.L. Sun, X.H. He, Tuning optical properties of photonic crystal of anodic alumina and the influence of electrodeposition. J. Electrochem. Soc. **156**(11), D521–D524 (2009)
3. P.J. Stout, D. Zhang, High-Power magnetron Cu seed deposition on 3-D dual inlaid features. IEEE Trans. Plasma Sci. **30**(1), 116–117 (2002)
4. W.M. Zhang, M.H. Deng, Y.Q. Pei, Z.A. Wang, Design and optimization of high current power supply for electrochemistry, in *Power Electronics Conference IPEC*, (2010), pp. 86–91
5. G. Zhang, Z. Li, B. Zhang, D. Qiu, W. Xiao, W.A. Halang, A Z-Source Half-Bridge Converter. IEEE Trans. Ind. Electron. **61**(3), 1269–1279 (Mar. 2014)
6. X.P. Ding, Z.M. Qian, S.T. Yang, B. Cuil, F.Z. Peng, A direct peak DC-link boost voltage control strategy in Z-source inverter, in *22nd Annual IEEE Applied Power Electronics Conference (APEC 2007)*, (Mar. 2007) pp. 648–653

Chapter 8
Dual-Output Z-source Half-Bridge Converter

The design of a novel dual-output Z-source half-bridge converter with only three switches is to be presented in this chapter. In detail, two impedance networks are parallelized to reduce the output impedance. Therein, a few number of the capacitors and switches are reduced. Moreover, the proposed converter can realise buck-boost functions, which reduces the cost and increases the watt density.

8.1 Introduction

Under the pressure of the energy crisis and environmental pollution in the last decades, governments have been making the greatest efforts to seek new solutions [1, 2], one of which is to develop hybrid electric vehicles (short for HEV) [4], whose diagram is shown in Fig. 8.1, including a battery V_d (Batt.), a bidirectional DC-DC boost converter, DC buses (negative and positive), two DC-AC inverters, a generator (Gen.), an electric motor (Mot.), and an internal combustion engine (short for ICE). Therein, the DC-DC boost converter is used to increase the voltage of the battery to realize the DC bus voltage, while the DC bus offers or absorbs energy to/from two inverters. The interface between the DC bus and the electric motor is one of the inverters. The inverter operates as a DC-AC converter when the electric motor consumes power to drive the vehicle, while operates as an AC-DC converter in the deceleration or braking to save the energy. The other inverter is to connect the DC bus and the electric generator, and works as a DC-AC converter to start the ICE and be an AC-DC converter to absorb energy [3].

To reduce the number of switches, a dual-output converter with nine switches is proposed in [3]. Therein, two inverters share three common switches, resulting in the reduction of three switches. However, there are still the limited voltage and shoot-through problems. In order to solve these problems, the DC-DC boost converter in Fig. 8.1 is replaced with a Z-network in [3], in which 5 switches are reduced compared to the one in Fig. 8.1. However, there are still 9 switches in it.

© Springer International Publishing AG 2018

G. Zhang et al., *Designing Impedance Networks Converters*, Studies in Systems, Decision and Control 119, DOI 10.1007/978-3-319-63655-9_8

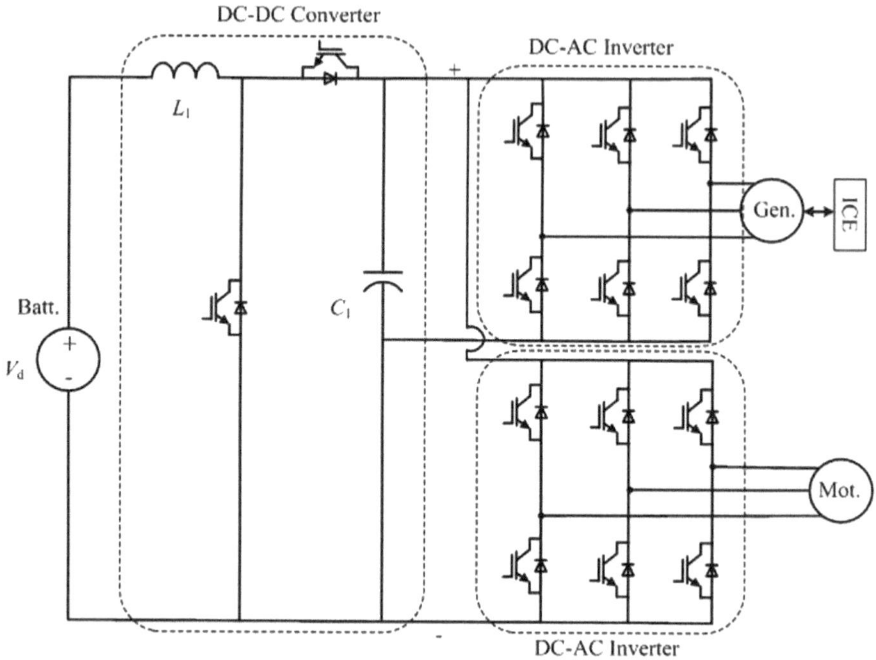

Fig. 8.1 Structure of conventional hybrid vehicle systems

Further to reduce the number of switches, a dual-output Z-source half-bridge converter with only three switches for hybrid electric vehicle systems is proposed [5] along with the impedance networks matching and the design methodology in this chapter.

8.2 Topology Design

As aforementioned above, electric vehicles required power supplies systems with dual-output DC-AC converters because there are two energy systems. Moreover, specific features of few switches, being immunity to the shoot-through problems, and high output voltage gains are eagerly required. Therefore, in terms of the impedance network matching and the designing methodology, the impedance networks matching is used to analyse the disadvantages of traditional one and improve it. Half-bridge DC-AC converters are chosen as the topology because the main purpose of this chapter is to reduce the number of switches and there are only two switches in it.

Since paralleled connectivity can reduce the impedance, paralleled connectivity is used to reduce the output impedance to realize output impedance matching. Moreover, in order to realize two output voltage, two half-bridge DC-AC converters are

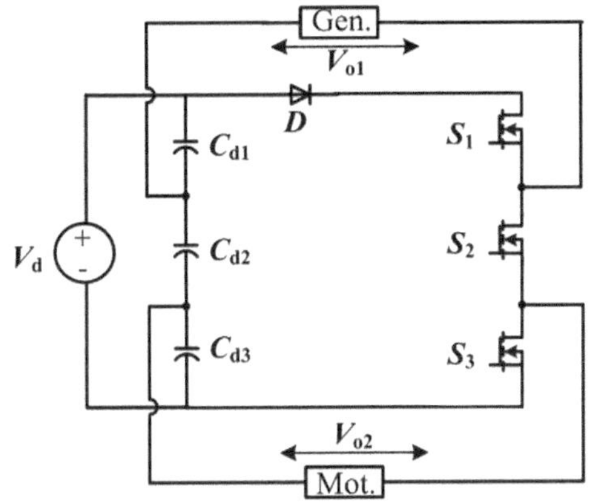

Fig. 8.2 A dual-output half-bridge converter with three switches

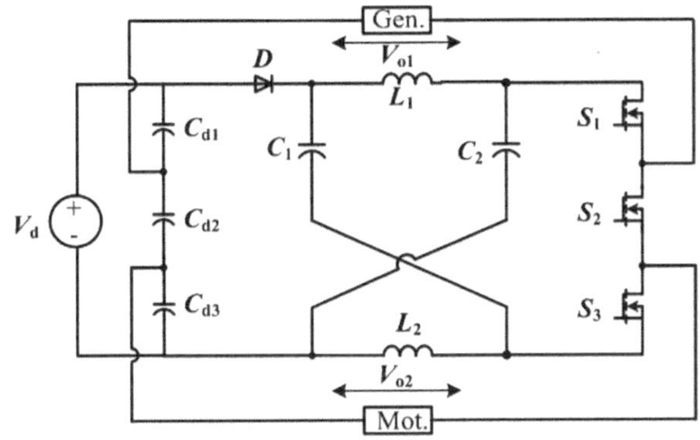

Fig. 8.3 A dual-output Z-source half-bridge converter with three switches

paralleled. In order to realize features of few switches and inspired from the nine switches converter, one capacitor and one switches are shared by two converters, which results in the novel converter as shown in Fig. 8.2. Further in order to realize features of being immunity to the shoot-through problems and high output voltage gains, an X-shaped two-port impedance network is embedded into the traditional half-bridge converter in terms of Table 5.3, which finally results in the novel converter as shown in Fig. 8.3.

The proposed converter is depicted in Fig. 8.3, in which two outputs share the same capacitor C_{d_2} and the same switch S_2. Then, the proposed converter own the same features as the features of two Z-source half-bridge converter in Fig. 7.3. Therein, capacitors C_{d_1}, C_{d_2}, switches S_1, S_2 offer a loop for the load Gen, while capacitors C_{d_2}, C_{d_3}, switches S_2, S_3 form another loop for the load Mot, and these two loops share the same Z-network, consisting of capacitors C_1 and C_2 and inductors L_1 and L_2. Due to the Z-network, the proposed converter well solve the shoot-through and limited voltage problems [5]. The unique feature of the proposed converter is that there are only three switches, which well reduces the economic cost and increase the efficiency of the electric vehicle.

8.3 Control Strategy

Corresponding to the on and off states of three switches, there are 8 kinds of modes, which result in different output results in the proposed converter. As shown in Fig. 8.4, the time sequence indicates the driving waveforms of the three switches.

According to Fig. 8.4, one can obtain different control strategy by adjusting the duties and phase-shift of three switches.

Therein, d_1, d_2 and d_3 are the duties of switches S_1, S_2 and S_3, respectively. Moreover, Δd_{12}, Δd_{23} and Δd_{31} are the phase shift ratios from S_1 to S_2, S_2 to S_3, and S_3 to S_1, respectively. Denote three binary coded digits B_1, B_2 and B_3 as the states of S_1, S_2 and S_3. $B_i = 1 (i = 1, 2, 3)$ means that the switch S_i is on and $B_i = 0$ implies that the switch S_i is off. Hence, $B_1 B_2 B_3$ corresponds to 8 modes, i.e. 000, 001, 010, 011, 100, 101, 110, and 111. One mode is introduced as a typical example.

For simplicity, assume that $\Delta d_{12} = \Delta d_{23} = \Delta d_{31} = \Delta d = \frac{1}{3}$, and $d_1 = d_2 = d_3 = d_c$. In order to avoid the saturations of inductors, the time interval of shoot-through states should be smaller than that of non-shoot-through states in the proposed converter, then one can obtain $d_c - 2\Delta d < 1 - d_c$ from Fig. 8.4, namely, $0 < d_c < \frac{5}{6}$. Therefore, there are five cases due to the conducted switches in each operational mode:

Case 1: one or zero conducted switch, namely, $0 < d_c < \frac{1}{3}$;
Case 2: one conducted switch, namely, $d_c = \frac{1}{3}$;
Case 3: two or one conducted switches, namely, $\frac{1}{3} < d_c < \frac{2}{3}$;
Case 4: two conducted switches, namely, $d_c = \frac{2}{3}$; and
Case 5: two or three conducted switches, namely, $\frac{2}{3} < d_c < \frac{5}{6}$, which are demonstrated in Fig. 8.5.

It is thus clear that Case 5 contains other four cases as special cases. Therefore, Case 5 is taken as an example.

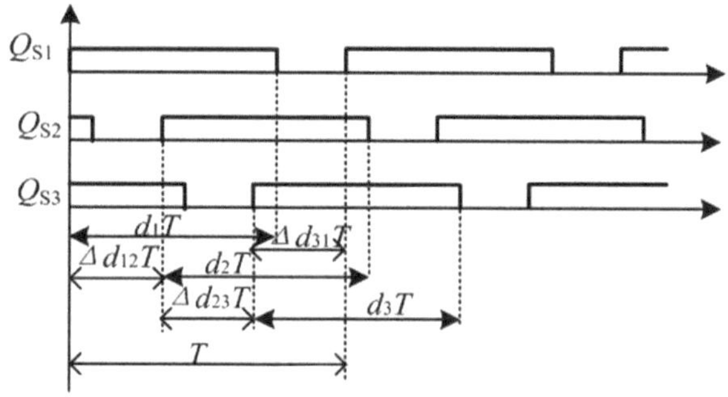

Fig. 8.4 Waveforms of the three switches

Fig. 8.5 Modes transition diagram in five different cases

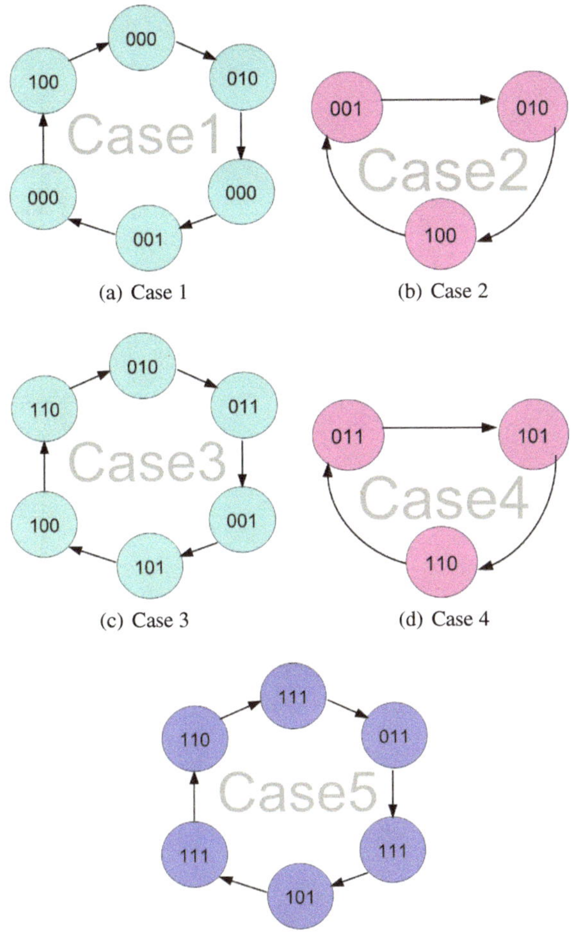

(a) Case 1 (b) Case 2 (c) Case 3 (d) Case 4 (e) Case 5

8.4 Operational Mode Analysis

For simplicity, it is assumed that (a) all the components are ideal; (b) the dead time in the driven pulses is ignored; (c) $L_1 = L_2$ and $C_1 = C_2$ in the Z-network; and (d) C_1, C_2, C_{d_1}, C_{d_2} and C_{d_3} are large enough. Then, assume the voltages of C_{d_1}, C_{d_2}, C_{d_3}, C_1, and C_2 to be constant in steady states, denoted by $V_{C_{d_1}}$, $V_{C_{d_2}}$, $V_{C_{d_3}}$, V_{C_1} and V_{C_2}, respectively.

Z_1 and Z_2 are used to represent loads Gen. and Mot. in Fig. 8.3, respectively.

There are four switching modes in Case 5 due to the behavior of the switches, which correspond to four equivalent circuits as shown in Fig. 8.6a, b, c, and d, respectively. Therein, the corresponding operational modes are: Mode 1, S_1, S_2 and S_3 are on, D is off ($B_1B_2B_3 = 111$); Mode 2, S_1, D and S_3 are on, S_2 is off ($B_1B_2B_3 = 101$);

Fig. 8.6 Equivalent circuits in Case 5

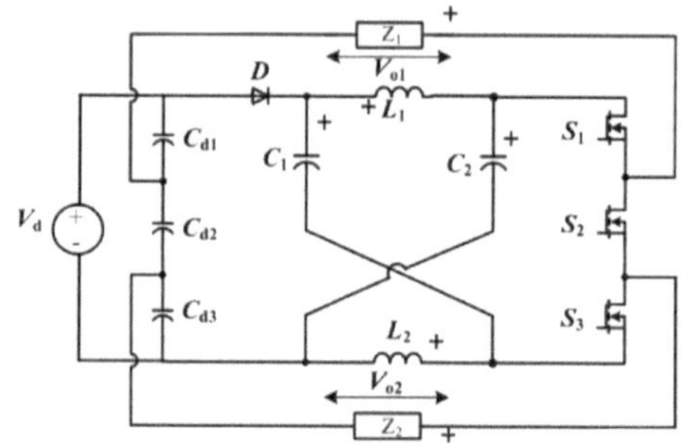

(a) Mode 1

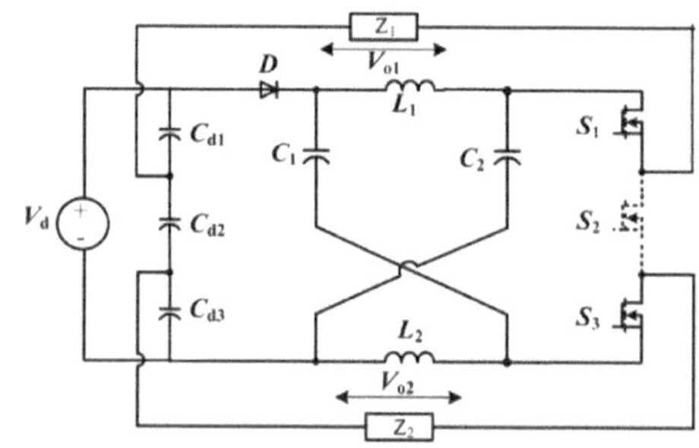

(b) Mode 2

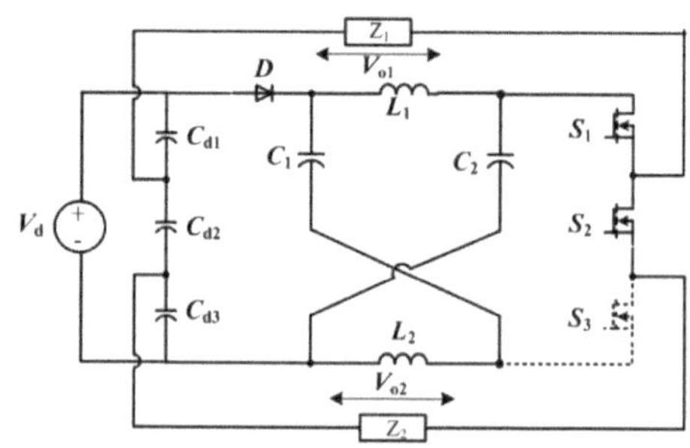

(c) Mode 3

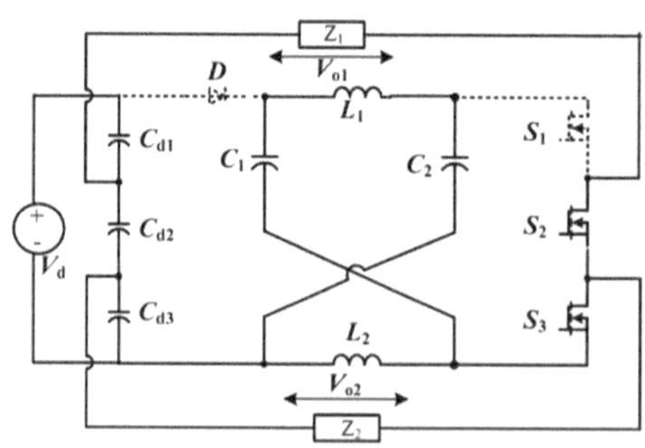

(d) Mode 4

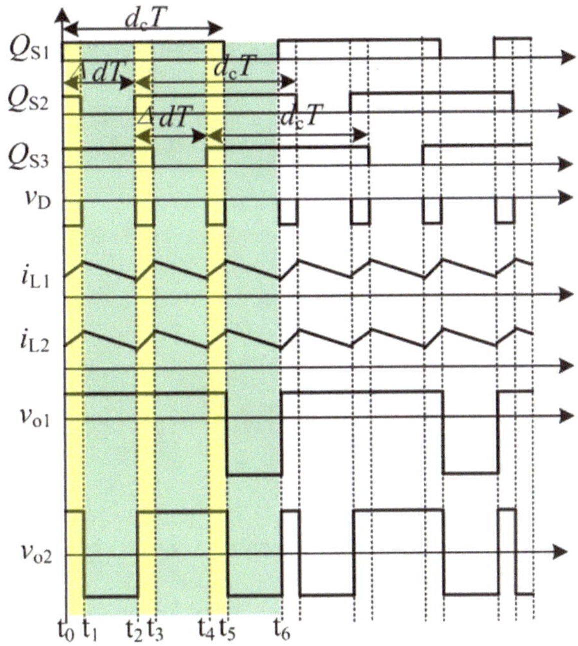

Fig. 8.7 Key waveforms of the dual-output Z-source half-bridge converter in Case 5: Shoot-through case

Mode 3, S_1, S_2 and D are on, S_3 is off ($B_1B_2B_3 = 110$); and Mode 4, when D, S_2 and S_3 are on, S_1 is off ($B_1B_2B_3 = 011$).

Denote t_0 the initial time of one period,
t_1 the mode transition instant from Mode 1 to Mode 2, i.e. $t_1 = t_0 + (d_c + \Delta d - 1)T$;
t_2 the mode transition instant from Mode 2 to Mode 1, i.e. $t_2 = t_0 + \Delta dT$;
t_3 the mode transition instant from Mode 1 to Mode 3, i.e. $t_3 = t_2 + (d_c + \Delta d - 1)T$;
t_4 the mode transition instant from Mode 3 to Mode 1, i.e. $t_4 = t_2 + \Delta dT$;
t_5 the mode transition instant from Mode 1 to Mode 4, i.e. $t_5 = t_4 + (d_c + \Delta d - 1)T$;
and $t_6 = T$ the end of the period.

In the steady state of the converter, its operational process in a switch period is analysed in the following and the key waveforms of Case 5 are shown in Fig. 8.7, which depicts the driven voltages of switches S_1, S_2, and S_3, namely, Q_{S1}, Q_{S2} and Q_{S3}; the voltage of diode D, namely v_D; the currents of inductors L_1 and L_2, namely, i_{L_1} and i_{L_2}; and the output voltages v_{o_1} and v_{o_2}, accordingly.

In the following, the output voltages are to be derived in each mode.

Mode 1$(t \in ([t_0, t_1] \cup [t_2, t_3] \cup [t_4, t_5]))$
As shown in Fig. 8.6a, the diode D is off and three switches are on. Then, the shoot-through state occurs. Therein, the source and Z-network transfer energy to Z_1 and Z_2. In detail, the source V_d and capacitor C_2 transfer energy to Z_1 via V_d-C_{d_1}-Z_1-S_1-C_2 loop, then the voltage of Z_1 is $v_{o_1} = V_{C_{d_1}} + V_{C_2} - V_d$. Moreover, C_{d_3} and C_2 discharge energy to Z_2 through C_{d_3}-C_2-S_1-S_2-Z_2 loop, namely, $v_{o_2} = V_{C_2} - V_{C_{d_3}}$. Due to the shoot-through state, the voltages of L_1 and L_2 are equal to the voltages

of capacitors C_1 and C_2, namely $v_{L_1} = v_{L_2} = V_{C_1} = V_{C_2}$. This mode lasts so long as $(d_c + \Delta d - 1)T$. Therefore, one can obtain the output voltages as

$$\begin{cases} v_{o_1} = V_{C_{d_1}} + V_{C_2} - V_d \,, \\ v_{o_2} = V_{C_2} - V_{C_{d_3}} \,. \end{cases} \tag{8.1}$$

Mode 2 $(t \in [t_1, t_2])$
As shown in Fig. 8.6b, D, S_1 and S_3 are on, S_2 is off. Therein, V_d transfers energy to Z-network through D, hence, one can obtain that $v_{L_1} = V_d - V_{C_2}$. In the V_d-C_{d_1}-Z_1-S_1-C_2 loop, V_d and C_2 discharge energy to Z_1, namely, $v_{o_1} = V_{C_{d_1}} + V_{C_2} - V_d$. Moreover, Z_2 obtains energy from the V_d-D-C_1-S_3-Z_2-C_{d_3} loop, then, $v_{o_2} = V_d - V_{C_1} - V_{C_{d_3}} = V_d - V_{C_2} - V_{C_{d_3}}$. This mode lasts $(1 - d_c)T$. Then, one can obtain the output voltages as

$$\begin{cases} v_{o_1} = V_{C_{d_1}} + V_{C_2} - V_d \,, \\ v_{o_2} = V_d - V_{C_2} - V_{C_{d_3}} \,. \end{cases} \tag{8.2}$$

Mode 3 $(t \in [t_3, t_4])$
In Fig. 8.6c, D, S_1 and S_2 are on, S_3 is off. V_d transfers energy to Z-network through D, then one can obtain that $v_{L_1} = V_d - V_{C_2}$. Z_1 obtains energy as Modes 1 and 2. Meanwhile, Z_2 receives energy according to C_2-S_1-S_2-Z_2-C_{d_3} loop. This mode lasts $(1 - d_c)T$. Hence, the output voltages are given by

$$\begin{cases} v_{o_1} = V_{C_{d_1}} + V_{C_2} - V_d \,, \\ v_{o_2} = V_{C_2} - V_{C_{d_3}} \,. \end{cases} \tag{8.3}$$

Mode 4 $(t \in [t_5, t_6])$
As shown in Fig. 8.6d, D, S_2 and S_3 are on, S_1 is off. Therein, the source V_d charges the Z-network through D, then one can obtain that $v_{L_1} = V_d - V_{C_2}$. Z_1 obtains energy in terms of the C_{d_1}-D-C_1-S_3-S_2-Z_1 loop, namely, $v_{o_1} = V_{C_{d_1}} - V_{C_1} = V_{C_{d_1}} - V_{C_2}$. Z_2 gets energy from the V_d-D-C_1-S_3-Z_2-C_{d_3} loop, then $v_{o_2} = V_d - V_{C_1} - V_{C_{d_3}} = V_d - V_{C_2} - V_{C_{d_3}}$. This mode lasts $(1 - d_c)T$. Therefore, one can obtain the output voltages as

$$\begin{cases} v_{o_1} = V_{C_{d_1}} - V_{C_2} \,, \\ v_{o_2} = V_d - V_{C_2} - V_{C_{d_3}} \,. \end{cases} \tag{8.4}$$

According to the analysis above, it concludes that

$$v_{o_1} = \begin{cases} V_{C_{d_1}} + V_{C_2} - V_d, & \text{for } t \in [t_0, t_5] \,, \\ V_{C_{d_1}} - V_{C_2}, & \text{for } t \in [t_5, t_6] \,, \end{cases} \tag{8.5}$$

$$v_{o_2} = \begin{cases} V_{C_2} - V_{C_{d_3}}, & \text{for } t \in \left([t_0, t_1] \bigcup [t_2, t_5]\right) \,, \\ V_d - V_{C_2} - V_{C_{d_3}}, & \text{for } t \in \left([t_1, t_2] \bigcup [t_5, t_6]\right) \,, \end{cases} \tag{8.6}$$

and

$$v_{L_1} = v_{L_2} = \begin{cases} V_{C_2}, & \text{for } t \in \left([t_0, t_1] \bigcup [t_2, t_3] \bigcup [t_4, t_5]\right) , \\ V_{\mathrm{d}} - V_{C_2}, & \text{for } t \in \left([t_1, t_2] \bigcup [t_3, t_4] \bigcup [t_5, t_6]\right) . \end{cases} \tag{8.7}$$

8.5 Deduction of Output Voltages

Based on the voltage-second characteristic of L_1, namely, $\int_0^T v_{L_1} dt = 0$, v_{L_1} can be derived as

$$\begin{aligned} &\int_0^T v_{L_1} dt \\ &= 3 \left(\int_0^{(d_{\mathrm{c}}+\Delta d-1)T} v_{L_1} dt + \int_{(d_{\mathrm{c}}+\Delta d-1)T}^{\Delta dT} v_{L_1} dt \right) \\ &= 3 \left(\int_0^{(d_{\mathrm{c}}+\Delta d-1)T} V_{C_2} dt + \int_{(d_{\mathrm{c}}+\Delta d-1)T}^{\Delta dT} (V_{\mathrm{d}} - V_{C_2}) dt \right) \\ &= 0 . \end{aligned} \tag{8.8}$$

Then, one obtains V_{C_1} and V_{C_2} as

$$V_{C_1} = V_{C_2} = \frac{1 - d_{\mathrm{c}}}{2 - (\Delta d + 2d_{\mathrm{c}})} V_{\mathrm{d}} . \tag{8.9}$$

Substituting (8.9) into (8.7) leads to

$$v_{L_1} = v_{L_2} = \begin{cases} \dfrac{1 - d_{\mathrm{c}}}{2 - (\Delta d + 2d_{\mathrm{c}})} V_{\mathrm{d}}, & \text{for } t \in \left([t_0, t_1] \bigcup [t_2, t_3] \bigcup [t_4, t_5]\right) , \\ \dfrac{1 - (\Delta d + d_{\mathrm{c}})}{2 - (\Delta d + 2d_{\mathrm{c}})} V_{\mathrm{d}}, & \text{for } t \in \left([t_1, t_2] \bigcup [t_3, t_4] \bigcup [t_5, t_6]\right) . \end{cases} \tag{8.10}$$

The ampere-second property of C_{d_1}, C_{d_2} and C_{d_3} implies

$$\begin{cases} \int_0^T i_{C_{d_2}} dt = 0 , \\ \int_0^T i_{C_{d_1}} dt = \int_0^T (i_{C_{d_2}} - i_{\mathrm{o}_1}) dt = \int_0^T i_{\mathrm{o}_1} dt = 0 , \\ \int_0^T i_{C_{d_3}} dt = \int_0^T (i_{C_{d_2}} + i_{\mathrm{o}_2}) dt = \int_0^T i_{\mathrm{o}_2} dt = 0 , \end{cases} \tag{8.11}$$

where $i_{C_{d_1}}$, $i_{C_{d_2}}$,$i_{C_{d_3}}$, i_{o_1} and i_{o_2} are the currents of C_{d_1}, C_{d_2}, C_{d_3}, Z_1 and Z_2, respectively.

Then, one obtains

$$\begin{cases} \int_0^T i_{o_1} dt = \int_0^T \frac{v_{o_1}}{Z_1} dt = 0\,, \\ \\ \int_0^T i_{o_2} dt = \int_0^T \frac{v_{o_2}}{Z_2} dt = 0\,. \end{cases} \tag{8.12}$$

Substituting (8.5) and (8.6) into (8.12) leads to

$$\begin{cases} \frac{V_{C_{d_1}} + V_{C_2} - V_{\mathrm{d}}}{Z_1} d_{\mathrm{c}} T + \frac{V_{C_{d_1}} - V_{C_2}}{Z_1} (1 - d_{\mathrm{c}}) T = 0\,, \\ \\ \frac{V_{C_2} - V_{C_{d_3}}}{Z_2} (2d_{\mathrm{c}} - 1) T + \frac{V_{\mathrm{d}} - V_{C_2} - V_{C_{d_3}}}{Z_2} (2 - 2d_{\mathrm{c}}) T = 0\,. \end{cases} \tag{8.13}$$

From (8.13), one obtains $V_{C_{d_1}}$ and $V_{C_{d_3}}$ as

$$\begin{cases} V_{C_{d_1}} = V_{C_2}(1 - 2d_{\mathrm{c}}) + V_{\mathrm{d}} d_{\mathrm{c}} = \frac{1 - d_{\mathrm{c}} - d_{\mathrm{c}} \Delta d}{2 - (\Delta d + 2d_{\mathrm{c}})} V_{\mathrm{d}}\,, \\ \\ V_{C_{d_3}} = V_{C_2}(4d_{\mathrm{c}} - 3) + V_{\mathrm{d}}(2 - 2d_{\mathrm{c}}) = \frac{(1 - d_{\mathrm{c}})(1 - 2\Delta d)}{2 - (\Delta d + 2d_{\mathrm{c}})} V_{\mathrm{d}}\,. \end{cases} \tag{8.14}$$

Substituting (8.14) and (8.9) into (8.5) and (8.6) leads to

$$v_{o_1} = \begin{cases} \frac{\Delta d - \Delta d d_{\mathrm{c}}}{2 - (\Delta d + 2d_{\mathrm{c}})} V_{\mathrm{d}}, \text{ for } t \in [t_0, t_5]\,, \\ \\ \frac{-\,\Delta d d_{\mathrm{c}}}{2 - (\Delta d + 2d_{\mathrm{c}})} V_{\mathrm{d}}, \text{ for } t \in [t_5, t_6]\,, \end{cases} \tag{8.15}$$

and

$$v_{o_2} = \begin{cases} \frac{2\Delta d(1 - d_{\mathrm{c}})}{2 - (\Delta d + 2d_{\mathrm{c}})} V_{\mathrm{d}}, \text{ for } t \in \left([t_0, t_1] \bigcup [t_2, t_5]\right)\,, \\ \\ \frac{\Delta d(1 - 2d_{\mathrm{c}})}{2 - (\Delta d + 2d_{\mathrm{c}})} V_{\mathrm{d}}, \text{ for } t \in \left([t_1, t_2] \bigcup [t_5, t_6]\right)\,. \end{cases} \tag{8.16}$$

Assume $\Delta d = 1/3$, then in terms of (8.15) and (8.16), one can obtain two output voltage gains M_1 and M_2 as

$$M_1 = \begin{cases} \dfrac{1-d_c}{5-6d_c}, & \text{for } t \in [t_0, t_5]\,, \\[2ex] \dfrac{-d_c}{5-6d_c}, & \text{for } t \in [t_5, t_6]\,, \end{cases} \tag{8.17}$$

and

$$v_{o_2} = \begin{cases} \dfrac{2-2d_c}{5-6d_c}, & \text{for } t \in \left([t_0, t_1] \bigcup [t_2, t_5]\right)\,, \\[2ex] \dfrac{1-2d_c}{5-6d_c}, & \text{for } t \in \left([t_1, t_2] \bigcup [t_5, t_6]\right)\,. \end{cases} \tag{8.18}$$

In order to make the proposed converter operate in Case 5, there is a constraint condition in (8.17) and (8.18), i.e. $2/3 < d_c < 5/6$. Then, one can obtain the relationships between them, which are shown in Fig. 8.8, and the zoomed-in figure is shown in Fig. 8.9.

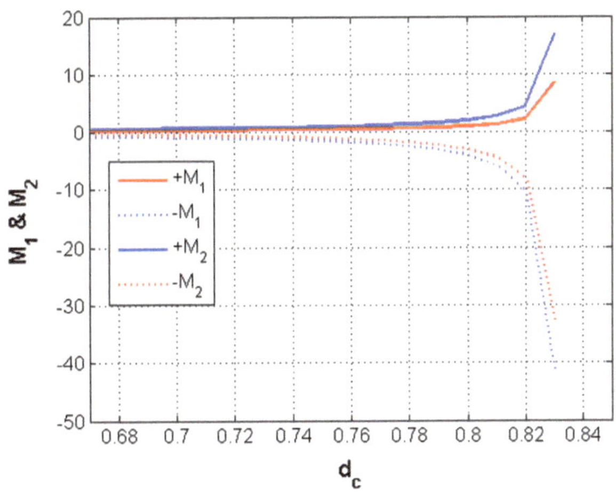

Fig. 8.8 Curves of dual output voltage gains

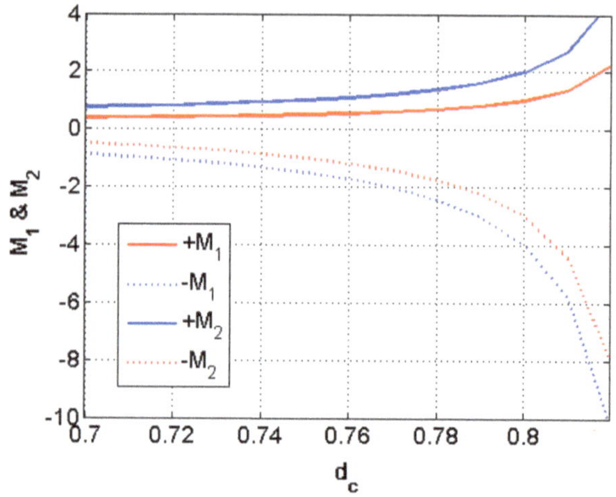

Fig. 8.9 Zooming in of Fig. 8.8

8.6 Parameters Determination

8.6.1 Parameters of Capacitors

Normally, the design of the capacitor is to determine the rated voltage and capacitance with a permitted fluctuation range, $x_C\%$ (x_C is pre-assigned), given output voltages V_{o_1} and V_{o_2}, output currents I_{o_1} and I_{o_2}, and the switching period T.

8.6.1.1 Determination of the Rated Voltage

From (8.9), the rated voltage of C_2 can be determined by the maximal V_{C_2}, namely, $V_{C_2\text{M}}$.

8.6.1.2 Determination of the Rated Capacitance

The ripples of the capacitors have great influence on the stability of the converter, whose permitted fluctuation range can be used to determine the capacitance.

Then, the capacitors in Z-network can be determined according to the differential equation of capacitors

$$C_2 = \frac{i_{C_2}dt}{dv_{C_2}}. \tag{8.19}$$

The high harmonic frequency of the capacitance is nearly equal to triple to the switching frequency of the converter, namely

$$dt \approx \frac{T}{3}. \tag{8.20}$$

Denote the permitted error of V_{C_2M} by dv_{C_2}, according to the permitted fluctuation range $x_C\%$, dv_{C_2} is expressed as

$$dv_{C_2} = x_C\% V_{C_2\mathrm{M}}\,. \tag{8.21}$$

Substituting (8.20) and (8.21) into (8.19) leads to

$$C_2 = \frac{I_{o_2}\frac{T}{3}}{x_C\% V_{C_2\mathrm{M}}}. \tag{8.22}$$

Therein, the range of the capacitance can be calculated, and the maximal one takes the rated capacitance.

8.6.2 *Parameters of Inductors*

Similarly, the parameter determination of the inductor is to determine the rated current and capacitance with a permitted fluctuation range $x_L\%$ (x_L is pre-assigned), given output voltages V_{o_1} and V_{o_2}, output currents I_{o_1} and I_{o_2}, and the switching period T.

8.6.2.1 Determination of the Rated Current

In fact, the rms current of L_2 is approximated to be I_{o_2}. Considering the safety margin, the rated current of L_2 is normally taken as $2I_{o_2}$.

8.6.2.2 Determination of the Rated Inductance

The ripples of the inductors also have great influence on the stability of the converter; therefore, the inductance can be determined in terms of the permitted ripples.

The inductances in the Z-network can be determined according to the differential equation of inductances as

$$L_2 = \frac{v_{L_2} dt_L}{di_{L_2}}. \tag{8.23}$$

In Z-networks, one can obtain the maximal value of v_{L_2} from (8.10), namely, $V_{L_2\mathrm{M}}$.

The high harmonic frequency of the inductance is nearly equal to triple to the switching frequency of the converter, so the time interval dt_L in (8.23) can be obtained as

$$dt_L \approx \frac{T}{3}. \tag{8.24}$$

Denote the permitted error of I_{L_2} by di_{L_2}. According to the permitted fluctuation range $x_L\%$, di_{L_2} is expressed as

$$di_{L_2} = x_L\% I_{L_2}. \tag{8.25}$$

Then, one has the inductance of L_2 as

$$L_2 = \frac{V_{L_2\mathrm{M}}\frac{T}{3}}{x_L\% I_{o_2}}. \tag{8.26}$$

8.7 Simulations and Experiments

To verify the feasibility and validity of the proposed converter, PSIM software is applied for the simulation of the converter. The pre-assigned parameters are set as follows: $x_C\% = 1\%$, $x_L\% = 10\%$, $V_\mathrm{d} = 30$ V, $C_{d_1} = C_{d_2} = C_{d_3} = 470$ μF, $Z_1 = Z_2 = 50\ \Omega$, $\Delta d = 1/3$, $d_\mathrm{c} = 0.75$ and $T = 100$ μs.

According to the settings, the parameters of the converter can be calculated: $C_1 = C_2 = 182.5$ μF, $L_1 = L_2 = 105.5$ μH, based on which the parameters can be practically selected as follows: $C_1 = C_2 = 220$ μF, $L_1 = L_2 = 100\ \mu H$.

The simulation results are shown in Fig. 8.10, which depicts the driven voltages of switches S_1, S_2, and S_3; the voltage of diode D; the currents of inductors L_1 and L_2; and the output voltages v_{o_1} and v_{o_2}, accordingly. Moreover, the output voltages are given by

$$v_{o_1} = \begin{cases} 15\ V, & \text{for } t \in [t_0, t_5]\ , \\ -45\ V, & \text{for } t \in [t_5, t_6]\ , \end{cases} \tag{8.27}$$

and

$$v_{o_2} = \begin{cases} 30\ V, & \text{for } t \in \left([t_0, t_1] \bigcup [t_2, t_5]\right)\ , \\ -30\ V, & \text{for } t \in \left([t_1, t_2] \bigcup [t_5, t_6]\right)\ . \end{cases} \tag{8.28}$$

It is remarked that the simulation results are consistent with the results of the theoretical analysis. Moreover, the positive and negative voltages of two outputs can be regulated by the variations of Δd and d_c. Generally speaking, the positive and negative amplitudes of two output voltages, and the duties of these two outputs can be modulated by d_1, d_2, d_3, Δd_{12}, Δd_{23}, Δd_{31}, and f, which are the very results required in HEVs.

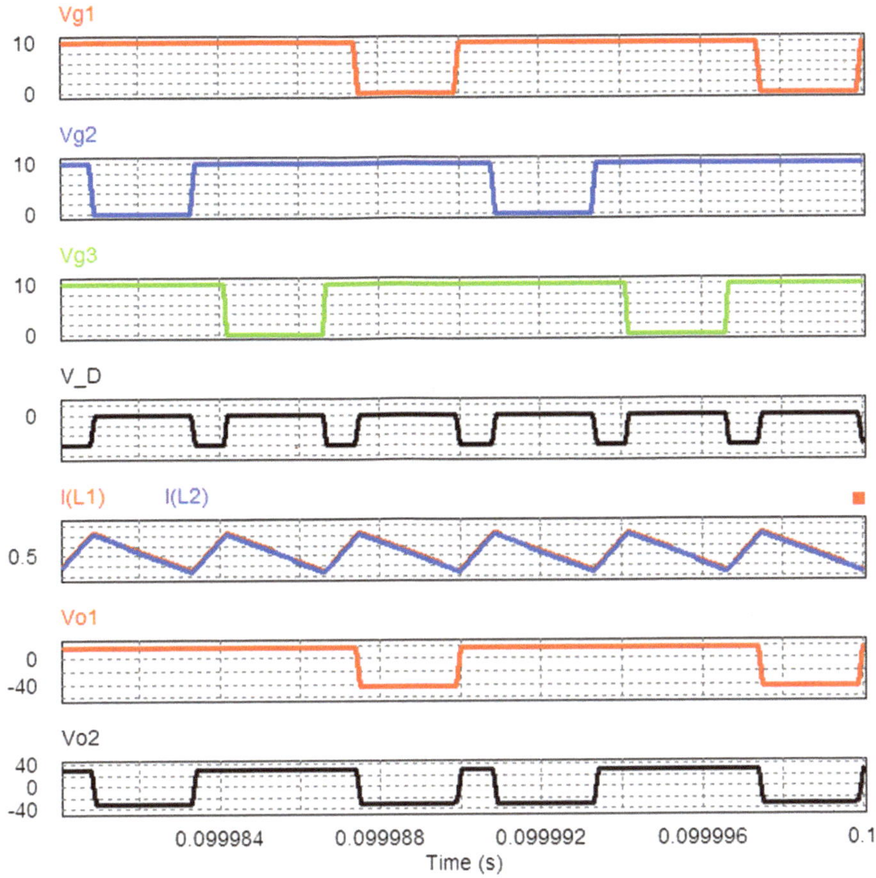

Fig. 8.10 Simulation waveforms of Case 5 when $\Delta d = \frac{1}{3}$ and $d_c = 0.75$

A prototype is then built as shown in Fig. 8.11, and the parameters are chosen as follow: $C_{d_1} = C_{d_2} = C_{d_3} = 470\ \mu F$, $C_1 = C_2 = 200\ \mu F$, $L_1 = L_2 = 100\ \mu H$, $Z_1 = Z_2 = 50\ \Omega$, and $T = 100\ \mu s$.

The prototype shown in Fig. 8.11 is composed of C_{d_1}, C_{d_2}, C_{d_3}, L_1, L_2, C_1, C_2, diode D (Type: MBRF20200), switches S_1, S_2, S_3 (Type: IRFP250A), and the resistive loads Z_1 and Z_2, with the driving circuits located in the left side of each switch. There three driving circuits are composed of three generated synchronous driving signals coming from RT-lab equipment (Type: OP5600), and driving ICs TLP2.

Figure 8.12 presents the experimental waveforms of the converter when $\Delta d = \frac{1}{3}$ and $d_c = 0.75$. Therein, Fig. 8.12 depicts the driven voltages of three switches, v_{g1}, v_{g2}, v_{g3}, the diode voltage v_D, the current of L_1, i_{L_1} and two output voltages, v_{o_1}, v_{o_2}, accordingly. In detail, the negative and positive output voltages of Z_1 are asymmetric. The positive one is about 15 V, and the negative one is nearly equal to

Fig. 8.11 Prototype of the proposed converter

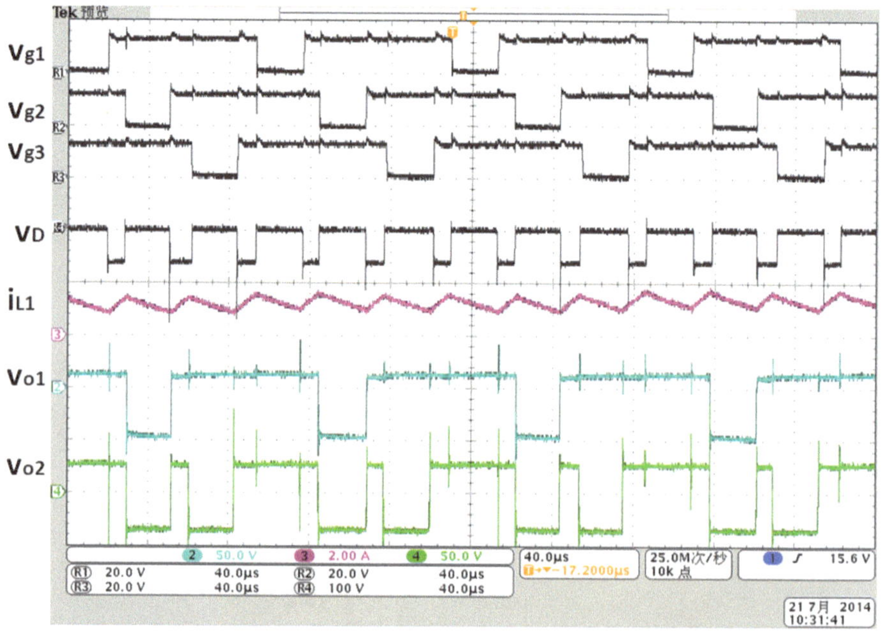

Fig. 8.12 Experimental waveforms in shoot-through case: Case 5

$V_d/2$, and the width is $d_c T$; while the negative one is -45 V, which is much larger than $V_d/2$. Moreover, the negative and positive output voltage of Z_2 are symmetric in amplitude, namely, ±30 V, and its width can be modulated. The experimental results are also consistent with the simulation results.

References

1. Z.Q. Zhu, D. Howe, Electrical machines and drives for electric, hybrid, and fuel cell vehicles. Proc. IEEE **95**(4), 746–765 (2007)
2. C.C. Chan, The state of the art of electric, hybrid, and fuel cell vehicles. Proc. IEEE **95**(4), 704–718 (2007)
3. S.M. Dehghan, M. Mohamadian, A. Yazdian Varjani, Hybrid electric vehicle based on bidirectional Z-source nine-switch inverter. IEEE Trans. Veh. Technol. **59**(6), 2641–2653 (2010)
4. J. Liu, H. Peng, Modeling and control of a power-split hybrid vehicle. IEEE Trans. Control. Syst. Technol. **16**(6), 1242–1251 (2008)
5. G. Zhang, B. Zhang, Z. Li, Y. Zhang, S.-Z. Chen, A novel single-input-dual-output impedance network converter. IEEE J. Emerg. Sel. Top. Power Electron. (2017). doi:10.1109/JESTPE.2017.2655510

Chapter 9
Conclusions

By briefly retrospecting the development of power electronics (converters) in the past century, the challenges faced by the modern power electronics for applications in renewable energy have been clearly sketched out and raise much higher requirements for converters. The newly proposed impedance source converters (or Z-source converters) are of great potential to cope with these challenges, which constitutes the subject of this book:

Motivations: A qualitative analysis reveals the problems of conventional converters, like shoot-through, open-circuit, and limited output-voltage/current gain, which greatly hinder wide applications of power electronics in industry, and explains, in turn, why impedance-source converters can overcome these problems, demonstrating great potential for industrial applications.

Design Methodology: A set of criteria for designing impedance-source converters has been established by taking input- and output-impedance matching, and load-phase matching into account, which lead to a systematic design methodology instead of the traditional tedious, manual and experience-dependent design methods.

Applications: With this design methodology, several typical examples are presented in this book, i.e. two novel 3-Z-network DC-DC boost converters for solar energy systems, a Z-source half-bridge converter for electroplating applications and a dual-output Z-source half-bridge converter for hybrid electric vehicles.

This book can serve as a teaching material for graduate students or guidelines for engineers.

© Springer International Publishing AG 2018

G. Zhang et al., *Designing Impedance Networks Converters*, Studies in Systems, Decision and Control 119, DOI 10.1007/978-3-319-63655-9_9

FSC
www.fsc.org
MIX
Papier | Fördert
gute Waldnutzung
FSC® C083411

Kolibri 360 GmbH
Ferdinand-Jühlke-Straße 7
99095 Erfurt, Deutschland
produktsicherheit@kolibri360.de